POWER SYSTEM ANALYSIS AND DESIGN SOFTWARE

J. Duncan Glover

Associate Professor of Electrical and Computer Engineering
Northeastern University, Boston MA

PWS-KENT Publishing Company
Boston

PWS-KENT
Publishing Company

20 Park Plaza
Boston, Massachusetts 02116

PWS-KENT Publishing Company is a division of Wadsworth, Inc.

Printed in the United States of America
    90 91 -- 10 9 8 7 6 5 4 3

CONTENTS

Preface

1. INTRODUCTION   1

   Disk Preparation   2

   Getting Started   4

   Data Files   5

   Saving Cases   6

   Helpful Hints   8

2. MATRIX OPERATIONS (CHAP2)   10

3. SYMMETRICAL COMPONENTS (CHAP3)   13

4. LINE CONSTANTS (CHAP5)   16

5. TRANSMISSION LINES :

   STEADY-STATE OPERATION (CHAP6)   25

6. POWER FLOW (CHAP7)   29

7. SYMMETRICAL SHORT CIRCUITS (CHAP8)   42

8. SHORT CIRCUITS (CHAP9)   55

9. TRANSMISSION LINE TRANSIENTS (CHAP11)   66

10. TRANSIENT STABILITY (CHAP12)   74

11. PROGRAM LISTINGS AND FLOWCHARTS   83

12. SAMPLE SINGLE-LINE DIAGRAMS   97

PREFACE

Digital computer programs are used extensively in the analysis and design of electric power systems. With the advent of the personal computer, the user now has the convenience of access at any time or place. This software package contains a set of programs that enable engineering students and engineers to conduct power system analysis and design studies. The data-handling and number-crunching capabilities of the modern personal computer allow students and professionals to work on more difficult and realistic problems and make the PC an innovative tool in the learning process.

The floppy disks included with this software package can be run on the IBM PC, XT, AT or any compatible personal computer. The set of programs stored on the disks can be used to solve the personal computer problems given in the accompanying text *Power System Analysis and Design With Personal Computer Applications* , by J. D. Glover and M. Sarma , published by PWS Publishers in 1987. The user can also run the programs for his or her own power system studies.

This software package is primarily intended for education, not for solving large dimensional power systems. Thus the power flow, short circuit, and transient stability programs are capable of handling up to 30-bus power systems on a PC with 256K of memory, and up to 100-bus power systems on a PC with 1 Meg of memory.

Each of the following students wrote one of the programs in this software package, under the supervision of the author.

Jalil Elmernissi   -   LINE CONSTANTS (CHAP5)

Jean Y. Ayoub   -   POWER FLOW (CHAP7)

Terry J. Fundak   -   SYMMETRICAL SHORT CIRCUITS (CHAP8)

Mr. Fundak also assisted in the writing of SHORT CIRCUITS (CHAP9). The author is grateful to these students for their help in the preparation of the above programs.

The author hopes that this software package will be useful to students and professionals in their daily studies of power system analysis and design. For those who are studying from the text *Power System Analysis and Design With Personal Computer Applications* , the software can solve practical, real-world problems to supplement the theory, and can open the door to design orientation in power systems.

J. Duncan Glover

ECE Department

Northeastern University

Boston, MA 02115

1. INTRODUCTION

This manual gives user instructions for the computer programs stored on the floppy disks that accompany this software package. No programming knowledge is required on the part of the user to run these programs.

The programs are written to run on an IBM PC, XT, AT[*] or any compatible personal computer. The basic requirements are that the computer should have two 360K floppy disk drives, or one floppy disk and one hard disk drive, 256K of RAM, and a monochrome monitor. Additionally, if either the power flow, short circuit, or transient stability program is used for a power system with greater than 30 buses, more RAM is required (1 Meg for 100 buses). A printer is also needed for outputs.

Three disks entitled DISK 1, DISK 2, and DISK 3 accompany this software package. DISK 1 contains four programs: CHAP2, CHAP3, CHAP5, and CHAP6. DISK 2 contains three programs CHAP8, CHAP9, and CHAP11. Disk 3 contains two programs, CHAP7 and CHAP12. All of these programs are described in Chapters 2 - 10 of this manual, including input/output descriptions and sample runs. Program flowcharts are given in Chapter 11, and sample single-line diagrams for power system analysis and design studies are given in Chapter 12. The background and theory for the programs are given in the text *Power System Analysis and Design with Personal Computer Applications* , authored by J. D. Glover and M. Sarma, and published by PWS Publishers, Boston, 1987.

----

* IBM trademark

Disk Preparation

The three disks that accompany this software package, DISK 1, DISK 2, and DISK 3 , are write-protected (notchless), in order that you have a copy of the programs that cannot be easily erased or unintentionally modified. It is first necessary to copy the disks, since the computer will be writing into data files when you run some of the programs. You can store the original disks and use your copies for running programs.

TWO FLOPPY DISK DRIVES:

If your computer has two floppy disk drives, insert your own system disk in drive A, turn the computer power switch on, and boot the computer. Then format a 360K double-sided blank disk with a copy of DOS (version 2.0 or better) as follows. Insert a blank disk in drive B and keep your system disk in drive A. With the sign A> showing on the monitor, type:

FORMAT B:/S   < CR >

where < CR > indicates carriage return on the keyboard.

Next copy the contents of DISK 1 as follows. Remove your system disk and insert DISK1 in drive A. Keep the blank disk in drive B. Then with the sign A> showing on the monitor, type:

COPY *.* B: < CR >

Next copy DISK 2 and DISK 3 onto a second and third blank disk by repeating the above procedure. That is, format a second and third blank disk with DOS then copy the contents of DISK 2 and DISK 3 onto these disks. When you have completed this task, store the original disks away. The copies are now ready for you to execute the programs.

ONE FLOPPY DISK DRIVE AND ONE HARD DISK DRIVE :

If you have a hard disk, copy DISK 1, DISK 2, and DISK 3 onto the hard disk as follows. Insert DISK 1 in drive A. With the sign A> showing on the monitor, type:

COPY *.* C: < CR >

where < CR > indicates carriage return on the keyboard. Repeat with DISK 2 and then DISK3 in drive A. When you have completed this task, store the original disks away. Once the programs are copied onto the hard disk, they are ready to be executed.

TWO FLOPPY DISK DRIVES :

After you have formatted three blank disks with DOS and copied the contents of DISK 1, DISK 2, and DISK 3 on them, you are ready to execute the programs. Store away the original disks and work with the copies.

If you want to run CHAP2, CHAP3, CHAP5, or CHAP6, insert DISK 1 into drive A. If you want to run CHAP8, CHAP9, or CHAP11, insert DISK 2 into drive A. If you want to run CHAP7 or CHAP12, insert DISK 3 into drive A.

Suppose you want to execute CHAP2. With DISK 1 in drive A, turn the computer power switch on and boot the computer. Then with the A> sign showing on the monitor, type :

CHAP2 < CR >

where < CR > indicates carriage return. The program CHAP2 can now be executed. When you are finished with CHAP2, you can execute CHAP3, CHAP5, or CHAP6 in the same manner. Similarly, CHAP8, CHAP9, or CHAP11 can be executed by inserting DISK2 in drive A, and CHAP7 or CHAP12 can be executed by inserting DISK 3 in drive A.

ONE FLOPPY DISK DRIVE AND ONE HARD DISK :

After you have copied the contents of DISK 1, DISK 2, and DISK 3 onto the hard disk, you are ready to execute the programs. Store the original disks away and work from the hard disk.

Suppose you want to execute CHAP2. Turn the power switch on and boot the computer. Then with the C> sign showing on the monitor, type:

CHAP2   < CR >

The program CHAP2 can now be executed. When you are finished with CHAP2, you can execute CHAP3, CHAP5, CHAP6, CHAP7, CHAP8, CHAP9, CHAP11, OR CHAP12 in the same manner.

Data Files

DISK 1, DISK 2, and DISK 3 also contain data files. The data files are used with the CHAP5, CHAP7, CHAP8, CHAP9, CHAP11, and CHAP12 programs. All the data files have the extension " .DAT ".

Each program that uses data files has a "display the data file" selection. Therefore, the program user can display data either on the monitor or with the printer while executing the program. It is not necessary to directly examine the contents of a data file.

For each program that uses data files, the program user can store the data for three to five separate cases. For example, the data file that is used to store input data for Case 1 of CHAP5, LINE CONSTANTS, is called LC1.DAT. The input data file for Case 2 (3,4,5) has a similar name, except the number 2 (3,4,5) is used. Thus you can store the CHAP5 input data for up to five different transmission lines.

Each time you update the input data files for a program, the program writes over the data files. That is, it replaces the data files on your disk with the updated ones. In this way, the total disk space used for storage does not increase. Also, you don't have to erase any data files.

Saving Cases

For each program that uses data files, you can save three to five cases on a disk. If you have more cases that you want to save, then you can copy the disk. Or if you want a backup copy of the data for a particular case, you can copy the disk.

Consider for example, the program CHAP7, POWER FLOW. Suppose you have already executed CHAP7 for a particular 1990 power system consisting of 40 buses, 70 transmission lines and 20 transformers. Your data files for this study are stored as Case 1 with case name MIAMI EDISON-1990. You now want a backup copy of the MIAMI EDISON-1990 input/output data before studying system changes for the year 1995.

If you have two floppy disk drives, you can make a backup copy of the input/output data for MIAMI EDISON-1990 by copying the disk you are working with, DISK 3, onto a blank disk. With DISK 3 in drive A and the sign A> showing on the monitor, type:

DISKCOPY A: B: < CR >

Then insert a blank disk in drive B and follow the computer's instructions.

After you have copied the disk, store the copy away. You may want to make this copy write-protected by covering the notch on the disk with opaque tape.

If you have one floppy disk drive and a hard disk, you can make a backup copy of MIAMI EDISON-1990 by copying the input data files PFB1.DAT, PFL1.DAT, and PFT1.DAT and the output data files PFBO1.DAT, PFLO1.DAT, and PFTO1.DAT from your hard disk to a floppy disk. Then store away the floppy disk.

Helpful Hints

ENTERING DATA :

When entering data for any program on DISK 2 or DISK 3, it is helpful to have a single-line diagram of your power system directly in front of you. As such, you can see the system interconnections on the single-line diagram as you enter data. Also, all power system data should already be converted to per unit on a common system base.

ENTERING ZEROS :

To enter a 0 (zero) during data entry, you do not have to type in the number 0. Just press the < CR > key. It's faster this way.

CORRECTION OF ERROR BEFORE PRESSING < CR > :

Most of the programs give you options for changing the input data. However, if you make a mistake in data entry before pressing the < CR > key, the input data has not yet been registered in the computer. You can use the ← key to backspace and erase the mistake, then enter the correct data.

Ctrl PRINT SCREEN :

When it is time to print input/output data, the programs will tell you to use the Ctrl PRINT SCREEN option if you want to print the data on a printer. Press the Ctrl key and the PRINT SCREEN key simultaneously. Also, make sure your printer is turned on. The printer will then print out whatever appears on the monitor. You can use this option at any time.

2. MATRIX OPERATIONS (CHAP2)

This program contains the following nine subroutines :

(1) RMA - computes the matrix sum $C = A + B$ of the two N x M real matrices A and B.

(2) CMA - computes the matrix sum $C = A + B$ of the two N x M complex matrices A and B.

(3) RMM - computes the matrix product $C = A \times B$ of the N x M real matrix A and the M x P real matrix B.

(4) CMM - computes the matrix product $C = A \times B$ of the N x M complex matrix A and the M x P complex matrix B.

(5) RMI - computes the matrix inverse A^{-1} of the N x N real matrix A whose determinant is assumed to be nonzero.

(6) CMI - computes the matrix inverse A^{-1} of the N x N complex matrix A whose determinant is assumed to be nonzero.

(7) RMT - computes the matrix transpose A^{T} of the N x M real matrix A.

(8) CMT - computes the matrix transpose A^{T} of the N x M complex matrix A.

(9) CMC - computes the complex conjugate $A *$ of the N x M complex matrix A.

The maximum dimension (N, M, or P) of any matrix you enter is 30. You can enter the elements of complex matrices in either rectangular or polar form. The input/output data can be displayed on the monitor and the printer. Also, you can display the input/output data in either exponential format (similar to scientific notation) or fixed-point format (numbers with decimal points).

In the following sample run, the CMI subroutine is used to invert the 3 x 3 symmetrical components transformation matrix **A** given by (3.1.8) in the text *Power System Analysis and Design with Personal Computer Applications* , by J.D. Glover and M. Sarma.

SAMPLE RUN CHAP2 INVERTING A

```
THIS PROGRAM CONTAINS THE FOLLOWING NINE SUBROUTINES:

1:SUBROUTINE RMA (REAL MATRIX ADD)
2:SUBROUTINE CMA (COMPLEX MATRIX ADD)
3:SUBROUTINE RMM (REAL MATRIX MULTIPLY)
4:SUBROUTINE CMM (COMPLEX MATRIX MULTIPLY)
5:SUBROUTINE RMI (REAL MATRIX INVERSE)
6:SUBROUTINE CMI (COMPLEX MATRIX INVERSE)
7:SUBROUTINE RMT (REAL MATRIX TRANSPOSE)
8:SUBROUTINE CMT (COMPLEX MATRIX TRANSPOSE)
9:SUBROUTINE CMC (COMPLEX MATRIX CONJUGATE)


THE MAXIMUM DIMENSION OF ANY MATRIX YOU ENTER MUST BE 30 x 30 OR LESS

TO USE ONE OF THE ABOVE SUBROUTINES ENTER AN INTEGER FROM
1 TO 9. OR ENTER THE INTEGER 10 TO EXIT.

WHICH SUBROUTINE(ENTER AN INTEGER FROM 1 TO 10) ? 6


INPUT/OUTPUT DATA FOR SUBROUTINE CMI
WILL YOU ENTER THE DATA IN RECTANGULAR COORDINATES(Y OR N) ? n
DO YOU WANT THE OUTPUT DATA IN RECTANGULAR COORDINATES(Y OR N)? n
DO YOU WANT THE OUTPUT DATA IN EXPONENTIAL FORMAT(Y OR N)? y
WHAT IS THE DIMENSION N OF THE A MATRIX ? 3
```

FOR A COMPLEX MATRIX DENOTED A,
A = AMAG /_AANGLE WHERE AMAG IS THE MAGNITUDE
 AND AANGLE IS THE ANGLE(IN DEGREES) OF A.
IN OTHER WORDS ENTER A IN POLAR COORDINATES.

ENTER THE FIRST ROW OF THE MATRIX AMAG
ENTER EACH NUMBER THEN PRESS RETURN.
? 1 ? 1 ? 1
ENTER THE NEXT ROW OF THE MATRIX AMAG
? 1 ? 1 ? 1
ENTER THE NEXT ROW OF THE MATRIX AMAG
? 1 ? 1 ? 1
ENTER THE FIRST ROW OF THE MATRIX AANGLE
ENTER EACH NUMBER THEN PRESS RETURN.
? 0 ? 0 ? 0
ENTER THE NEXT ROW OF THE MATRIX AANGLE
? 0 ? 240   ? 120
ENTER THE NEXT ROW OF THE MATRIX AANGLE
? 0 ? 120   ? 240
USE THE Ctrl PRINT SCREEN OPTION NOW IF YOU WANT TO PRINT THE RESULTS
PRESS RETURN TO CONTINUE
THE MATRIX   A = AMAG /_ AANGLE IS

 1.000E+00 0.000E+00DEG 1.000E+00 0.000E+00DEG 1.000E+00 0.000E+00DEG
 1.000E+00 0.000E+00DEG 1.000E+00 2.400E+02DEG 1.000E+00 1.200E+02DEG
 1.000E+00 0.000E+00DEG 1.000E+00 1.200E+02DEG 1.000E+00 2.400E+02DEG

THE MATRIX   INVERSE AI = AIMAG /_ AIANGLE IS

 3.333E-01 5.123E-06DEG 3.333E-01-5.123E-06DEG 3.333E-01 0.000E+00DEG
 3.333E-01-2.561E-06DEG 3.333E-01 1.200E+02DEG 3.333E-01 2.400E+02DEG
 3.333E-01 0.000E+00DEG 3.333E-01 2.400E+02DEG 3.333E-01 1.200E+02DEG

REMOVE Ctrl PRINT SCREEN AND PRESS RETURN TO CONTINUE
THIS PROGRAM CONTAINS THE FOLLOWING NINE SUBROUTINES:

1:SUBROUTINE RMA (REAL MATRIX ADD)
2:SUBROUTINE CMA (COMPLEX MATRIX ADD)
3:SUBROUTINE RMM (REAL MATRIX MULTIPLY)
4:SUBROUTINE CMM (COMPLEX MATRIX MULTIPLY)
5:SUBROUTINE RMI (REAL MATRIX INVERSE)
6:SUBROUTINE CMI (COMPLEX MATRIX INVERSE)
7:SUBROUTINE RMT (REAL MATRIX TRANSPOSE)
8:SUBROUTINE CMT (COMPLEX MATRIX TRANSPOSE)
9:SUBROUTINE CMC (COMPLEX MATRIX CONJUGATE)

THE MAXIMUM DIMENSION OF ANY MATRIX YOU ENTER MUST BE 30 x 30 OR LESS

TO USE ONE OF THE ABOVE SUBROUTINES ENTER AN INTEGER FROM
1 TO 9. OR ENTER THE INTEGER 10 TO EXIT.

WHICH SUBROUTINE(ENTER AN INTEGER FROM 1 TO 10) ? 10

3.   SYMMETRICAL COMPONENTS (CHAP3)

This program contains the following three symmetrical component subroutines:

(1)   SEQVEC   -   computes the complex sequence vector $V_s = A^{-1} V_p$
for any three-phase complex phase vector V_p.

(2)   PHAVEC   -   computes the complex phase vector $V_p = A V_s$
for any three-phase complex sequence vector V_s.

(3)   SEQIMP   -   computes the 3 x 3 complex sequence impedance matrix
$Z_s = A^{-1} Z_P A$ for any 3 x 3 complex phase impedance
matrix Z_p.

The **A** matrix in the above subroutines is the 3 x 3 symmetrical components transformation matrix given by (3.1.8) of the text *Power System Analysis and Design With Personal Computer Applications*. Also, the **V** vectors in the above subroutines can be three-phase voltage vectors, three-phase current vectors, or any vectors with three complex components.

You can enter the elements of the complex vectors and matrices in either polar or rectangular form. The input/output data can be displayed on the monitor and the printer. Also, you can display the input/output data in either exponential format (similar to scientific notation) or in fixed point format (numbers with decimal points).

The following sample run computes the sequence components of an unbalanced three-phase voltage vector.

SAMPLE RUN   CHAP3   SEQUENCE COMPONENTS OF V_P

```
THIS PROGRAM CONTAINS THE FOLLOWING THREE SUBROUTINES:

1: SUBROUTINE SEQVEC(VP,VS)
     SEQVEC COMPUTES THE SEQUENCE VECTOR VS=(AINV)VP
2: SUBROUTINE PHASVEC(VP,VS)
     PHAVEC COMPUTES THE PHASE VECTOR VP=(A)VS
3: SUBROUTINE SEQIMP(ZP,ZS)
     SEQIMP COMPUTES THE SEQUENCE IMPEDANCE MATRIX ZS=(AINV)ZP(A)

 TO USE ONE OF THE ABOVE SUBROUTINES, ENTER AN INTEGER FROM
 1 TO 3 OR ENTER THE INTEGER 4 TO EXIT.

WHICH SUBROUTINE(ENTER AN INTEGER FROM 1 TO 4)? 1


WILL YOU ENTER THE DATA IN RECTANGULAR COORDINATES (Y OR N)? n

DO YOU WANT THE OUTPUT IN RECTANGULAR COORDINATES(Y OR N)? n

DO YOU WANT THE OUTPUT DATA IN EXPONENTIAL FORMAT (Y OR N)? n
```

```
*****INPUT/OUTPUT DATA FOR SUBROUTINE SEQVEC*****
THE COMPLEX VECTOR V IS DENOTED
V = VMAG /_ VANGLE,WHERE VMAG IS THE MAGNITUDE
AND VANGLE IS THE ANGLE(IN DEGREES) OF V.
IN OTHER WORDS, ENTER V IN POLAR COORDINATES.

ENTER THE FIRST ELEMENT OF THE VECTOR  VPMAG
? 200
ENTER THE SECOND ELEMENT OF THE VECTOR  VPMAG
? 190
ENTER THE THIRD ELEMENT OF THE VECTOR  VPMAG
? 215
THE COMPLEX VECTOR V IS DENOTED
V = VMAG /_ VANGLE,WHERE VMAG IS THE MAGNITUDE
AND VANGLE IS THE ANGLE(IN DEGREES) OF V.
IN OTHER WORDS, ENTER V IN POLAR COORDINATES.

ENTER THE FIRST ELEMENT OF THE VECTOR  VPANGLE
? 0
ENTER THE SECOND ELEMENT OF THE VECTOR  VPANGLE
? 250
ENTER THE THIRD ELEMENT OF THE VECTOR  VPANGLE
? 110
USE THE Cntrl PRINT SCREEN OPTION NOW IF YOU WANT TO PRINT THE RESULTS
PRESS RETURN TO CONTINUE
THE MATRIX  VP = VPMAG /_ VPANGLE IS

  200    0 DEG
  190    250 DEG
  215    110 DEG


THE MATRIX  VS = VSMAG /_ VSANGLE IS

  21.93908   20.91195 DEG
  199.621   -.4153456 DEG
  21.09858   197.6118 DEG


WHICH SUBROUTINE(ENTER AN INTEGER FROM 1 TO 4)? 4
```

4.   LINE CONSTANTS (CHAP5)

This program computes the series impedance and shunt admittance matrices of overhead single- and double-circuit three-phase transmission lines. The program also computes the electric field strength at the surface of the phase conductors and a lateral profile of ground-level electric field strength.

If your computer has graphics capability, the program CHAP5 will plot the ground-level electric field profile. If you have two floppy disk drives, it is first necessary to copy the file GRAPHICS.COM from your system disk to DISK1. Insert your system disk in drive A and DISK 1 in drive B. Then with the sign A> showing on the monitor, type:

COPY GRAPHICS.COM B: < CR >

If your computer has graphics capability and you are working from a hard disk, the file GRAPHICS.COM should already be stored on the hard disk. If your computer does not have graphics capability, you can still execute CHAP5, but the ground-level electric field part of the output display will be missing.

When the program is run, the user selects one of the following: load the input data, update the input data, save the input data, run the program, display the output, and exit.

The user can store the input data for up to five separate cases in data files called LC1.DAT, LC2.DAT, LC3.DAT, LC4.DAT, and LC5.DAT. When load or save is selected, the user is asked to enter an integer from one to five. The program then loads the input data from or saves input data to the input data file that is selected.

When the user selects update, the input data is displayed on the monitor (for example, see the enclosed sample run). The top row of the input data consists of: the number of circuits NC, where NC=1 for single-circuit or 2 for double-circuit; earth resistivity RHO, Ωm; rated line voltage, kV line-to-line; frequency F, Hz; right-of-way width ROW, m; and the number of neutral wires N1 for circuit 1 and N2 for circuit 2, where N1 and N2 can be 0,1, or 2.

After the first row of data, there is a row of data for each phase and for each neutral, as follows: conductor resistance R, Ω/km; conductor GMR, cm; conductor outside diameter D, cm; horizontal position X from the center of the right-of-way and vertical position Y, m; number of bundled conductors Nb, where Nb= 1,2,3 or 4; and the bundle spacing d, cm.

When NC = 1 for a single-circuit, the input data for circuit 2 are not used when the program is run. Also, when N1 or N2 = 0 for no neutral wires, the input data for NEUTRAL 1 and NEUTRAL 2 are not used when the program is run. Similarly, when N1 or N2 =1, the input data for NEUTRAL 2 are not used.

To make a data change, select update. The program will then display the input data and ask the user "are these correct (Y/N)?" The user should respond by pressing N for no. Next press either the up, down, left, or right cursor on the keyboard until the cursor is positioned under the number to be changed. Then press the space bar, type the new number and press return. Repeat this procedure for each number to be changed. After all changes are made, press the ESC key. The user can then save this data, run the program, and display the outputs.

After selecting run and the program has completed its computations, select display to obtain the output data. The output data consist of: the series phase impedance matrix Z_p, series sequence impedance matrix Z_s, shunt phase admittance matrix Y_p, shunt sequence admittance matrix Y_s, conductor surface electric field strength, and (if your computer has graphics capability) a plot of the ground-level electric field strength from the center to the edge of the right-of-way. The output also includes impedance and admittance matrices for completely transposed lines.

The following sample run is for a 765 kV, 60 Hz, single-circuit three-phase line having four ACSR 954 kcmil conductors per bundle, with 45.7 cm between conductors in the bundle. The line has flat horizontal phase spacing with 13.7 m between adjacent phases and 23 m average line height. There are two Alumoweld 7 no. 8 neutral wires located 10.5 m vertically above and ± 11 m horizontally from the center phase. The earth resistivity is 100 Ω m and the right-of-way width is 300 m. This sample run is also given in Example 5.10 of the text *Power System Analysis and Design With Personal Computer Applications*.

```
                    L <oad input data
                    U <pdate input data
                    S <ave input data from Disk
                    R <un the program
                    D <isplay the output
                    E <xit



          Enter Function:  <L>
```

```
                   INPUT   DATA   TABLE   LC1.DAT
```

	NC =	RHO =	V =	F =	ROW =	N1 =	N2 =
	R	GMR	D	X	Y	Nb	d

```
CIRCUIT 1
PHASE a
PHASE b
PHASE c
NEUTRAL 1
NEUTRAL 2
```

```
CIRCUIT 2
PHASE a
PHASE b
PHASE c
NEUTRAL 1
NEUTRAL 2
```

```
         INPUT DATA TABLE # FROM 1 TO 5 [LC1.DAT]? 1
```

```
                    L <oad input data
                    U <pdate input data
                    S <ave input data from Disk
                    R <un the program
                    D <isplay the output
                    E <xit



          Enter Function:  <U>
```

	NC = 1	RHO = 100	V = 765	F = 60	ROW = 300	N1 = 2	N2 = 2
	R	GMR	D	X	Y	Nb	d
CIRCUIT 1							
PHASE a	.0701	1.229	3.038	-13.7	23	4	45.7
PHASE b	.0701	1.229	3.038	0	23	4	45.7
PHASE c	.0701	1.229	3.038	13.7	23	4	45.7
NEUTRAL 1	1.52	.0636	.998	-11	33.5	1	0
NEUTRAL 2	1.52	.0636	.998	11	33.5	1	0
CIRCUIT 2							
PHASE a	.0701	1.228	3.38	5	20	3	40
PHASE b	.0701	1.228	3.38	15	20	3	40
PHASE c	.0701	1.228	3.38	25	20	3	40
NEUTRAL 1	3.75	.0636	.998	8	29	1	0
NEUTRAL 2	3.75	.0636	.998	22	29	1	0

ARE THESE CORRECT (Y/N) ? Y

```
L <oad input data
U <pdate input data
S <ave input data from Disk
R <un the program
D <isplay the output
E <xit
```

Enter Function: ⟨R⟩

```
L <oad input data
U <pdate input data
S <ave input data from Disk
R <un the program
D <isplay the output
E <xit




Enter Function: <D>




                    ENTER YOUR CASE NAME SAMPLE RUN
```

```
:                                                                            :
:     SINGLE CIRCUIT                       SAMPLE RUN                         :
:                                                                            :
:     SERIES PHASE IMPEDANCE MATRIX      Zp    EQ. 5-7-19       ohms/km       :
:                                                                            :
LMMMMMMMMMMMMMMMMMMMMMMMMMMMMMMMMMMMMMMMMMMMMMMMMMMMMMMMMMMMMMMMMMMMMMMMMMMMMM9
:                                                                            :
: 0.2283E+00 + j 7.189E-01 0.2122E+00 + j 3.979E-01 0.2084E+00 + j 3.499E-01 :
:                                                                            :
: 0.2122E+00 + j 3.979E-01 0.2323E+00 + j 7.121E-01 0.2122E+00 + j 3.979E-01 :
:                                                                            :
: 0.2084E+00 + j 3.499E-01 0.2122E+00 + j 3.979E-01 0.2283E+00 + j 7.189E-01 :
:                                                                            :
HMMMMMMMMMMMMMMMMMMMMMMMMMMMMMMMMMMMMMMMMMMMMMMMMMMMMMMMMMMMMMMMMMMMMMMMMMMMMM<
```

```
:                                                                            :
:     SINGLE CIRCUIT                       SAMPLE RUN                         :
:                                                                            :
:     Zp FOR A COMPLETELY TRANSPOSED LINE    Zphat  EQ. 5-7-21  ohms/km       :
:                                                                            :
LMMMMMMMMMMMMMMMMMMMMMMMMMMMMMMMMMMMMMMMMMMMMMMMMMMMMMMMMMMMMMMMMMMMMMMMMMMMMMMM9
:                                                                            :
: 0.2296E+00 + j 7.166E-01 0.2109E+00 + j 3.819E-01 0.2109E+00 + j 3.819E-01 :
:                                                                            :
: 0.2109E+00 + j 3.819E-01 0.2296E+00 + j 7.166E-01 0.2109E+00 + j 3.819E-01 :
:                                                                            :
: 0.2109E+00 + j 3.819E-01 0.2109E+00 + j 3.819E-01 0.2296E+00 + j 7.166E-01 :
:                                                                            :
HMMMMMMMMMMMMMMMMMMMMMMMMMMMMMMMMMMMMMMMMMMMMMMMMMMMMMMMMMMMMMMMMMMMMMMMMMMMMMMM<
```

21

```
SINGLE CIRCUIT                    SAMPLE RUN

SERIES SEQUENCE IMPEDANCE MATRIX Zs    EQ. 5-7-25     ohms/km


0.6514E+00 + j 1.480E+00 0.1060E-01 - j 9.113E-03 -.1319E-01 - j 4.620E-03
-.1319E-01 - j 4.620E-03 0.1875E-01 + j 3.348E-01 -.2904E-01 + j 1.815E-02
0.1060E-01 - j 9.113E-03 0.3024E-01 + j 1.608E-02 0.1875E-01 + j 3.348E-01
```

```
SINGLE CIRCUIT                    SAMPLE RUN

Zs FOR A COMPLETELY TRANSPOSED LINE    Zshat  EQ. 5-7-27   ohms/km


0.6514E+00 + j 1.480E+00 0.0000E+00 + j 0.000E+00 0.0000E+00 + j 0.000E+00
0.0000E+00 + j 0.000E+00 0.1875E-01 + j 3.348E-01 0.0000E+00 + j 0.000E+00
0.0000E+00 + j 0.000E+00 0.0000E+00 + j 0.000E+00 0.1875E-01 + j 3.348E-01
```

```
SINGLE CIRCUIT                    SAMPLE RUN

SHUNT PHASE ADMITTANCE MATRIX     Yp  EQ. 5-11-16      s/km


+ j 4.898E-06  - j 9.777E-07  - j 2.541E-07
- j 9.777E-07  + j 5.083E-06  - j 9.777E-07
- j 2.541E-07  - j 9.777E-07  + j 4.898E-06
```

```
:                                                                              :
:       SINGLE CIRCUIT                          SAMPLE RUN                      :
:                                                                              :
:       Yp FOR A COMPLETELY TRANSPOSED LINE     Yphat EQ. 5-11-17  s/km          :
:                                                                              :
LMMMMMMMMMMMMMMMMMMMMMMMMMMMMMMMMMMMMMMMMMMMMMMMMMMMMMMMMMMMMMMMMMMMMMMMMMMMMMMMMM9
:                                                                              :
:   + j 4.960E-06  - j 7.365E-07  - j 7.365E-07                                 :
:                                                                              :
:   - j 7.365E-07  + j 4.960E-06  - j 7.365E-07                                 :
:                                                                              :
:   - j 7.365E-07  - j 7.365E-07  + j 4.960E-06                                 :
:                                                                              :
HMMMMMMMMMMMMMMMMMMMMMMMMMMMMMMMMMMMMMMMMMMMMMMMMMMMMMMMMMMMMMMMMMMMMMMMMMMMMMMMMM<
```

```
:                                                                              :
:        SINGLE CIRCUIT                          SAMPLE RUN                      :
:                                                                              :
:      SHUNT SEQUENCE ADMITTANCE MATRIX  Ys     EQ. 5-11-20        s/km          :
:                                                                              :
LMMMMMMMMMMMMMMMMMMMMMMMMMMMMMMMMMMMMMMMMMMMMMMMMMMMMMMMMMMMMMMMMMMMMMMMMMMMMMMMMM9
:                                                                              :
: 0.0000E+00 + j 3.487E-06 -.1555E-06 + j 8.977E-08 0.1555E-06 + j 8.977E-08     :
:                                                                              :
: 0.1555E-06 + j 8.977E-08 0.7802E-13 + j 5.696E-06 0.4711E-06 - j 2.720E-07     :
:                                                                              :
: -.1555E-06 + j 8.977E-08 -.4711E-06 - j 2.720E-07 -.7802E-13 + j 5.696E-06     :
:                                                                              :
HMMMMMMMMMMMMMMMMMMMMMMMMMMMMMMMMMMMMMMMMMMMMMMMMMMMMMMMMMMMMMMMMMMMMMMMMMMMMMMMMM<
```

```
:                                                                              :
:        SINGLE CIRCUIT                          SAMPLE RUN                      :
:                                                                              :
:      Ys FOR A COMPLETELY TRANSPOSED LINE     Yshat EQ. 5-11-23  s/km           :
:                                                                              :
LMMMMMMMMMMMMMMMMMMMMMMMMMMMMMMMMMMMMMMMMMMMMMMMMMMMMMMMMMMMMMMMMMMMMMMMMMMMMMMMMM9
:                                                                              :
: 0.0000E+00 + j 3.487E-06 0.0000E+00 + j 0.000E+00 0.0000E+00 + j 0.000E+00     :
:                                                                              :
: 0.0000E+00 + j 0.000E+00 0.0000E+00 + j 5.696E-06 0.0000E+00 + j 0.000E+00     :
:                                                                              :
: 0.0000E+00 + j 0.000E+00 0.0000E+00 + j 0.000E+00 0.0000E+00 + j 5.696E-06     :
:                                                                              :
HMMMMMMMMMMMMMMMMMMMMMMMMMMMMMMMMMMMMMMMMMMMMMMMMMMMMMMMMMMMMMMMMMMMMMMMMMMMMMMMMM<
```

CONDUCTOR SURFACE ELECTRIC FIELD STRENGTH   Eqs.5-12-1 to 5-12-5

PHASE: 1   .   20.61891   kV/cm
PHASE: 2   .   22.5181   kV/cm
PHASE: 3   .   20.61891   kV/cm

MAXIMUM ELECTRIC FIELD IS: 22.5181 kV/cm

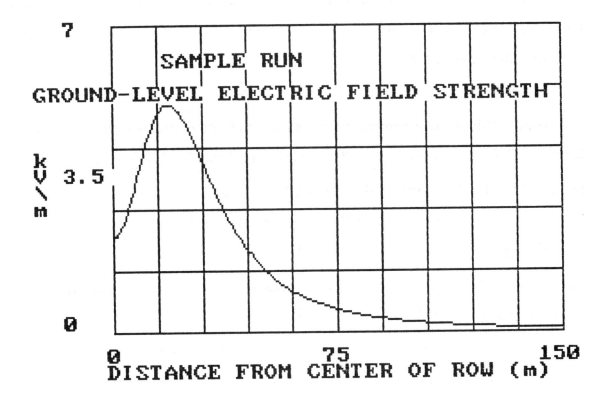

L <oad input data
U <pdate input data
S <ave input data from Disk
R <un the program
D <isplay the output
E <xit

Enter Function: <E>

5.   TRANSMISSION LINES - STEADY STATE OPERATION (CHAP6)

This program computes the ABCD parameters and equivalent π circuit values for a single-phase or completely transposed three-phase transmission line. The program also computes the sending end quantities: voltage; current; real, reactive, and apparent power; and power factor for a given receiving-end load, with or without compensation, assuming balanced steady-state operation. Voltage regulation and maximum power flow are also computed.

Input data to the program consist of : (1) rated line voltage and line length; (2) R, X_L, G, and B_C; (3) receiving-end full-load voltage, apparent power, and power factor; (4) the number (0, 1, or 2) and location of intermediate substations; and (5) the percent shunt reactive (inductive) and series capacitive compensation at each line terminal and at each intermediate substation. It is assumed that the shunt reactors are removed at heavy loads.

The output data consist of (a) without compensation: (1) characteristic impedance, (2) propagation constant, (3) wavelength, (4) surge impedance loading, (5) equivalent π circuit series impedance and shunt admittance, and (6) ABCD parameters; and (b) with compensation: (1) equivalent ABCD parameters, (2) sending-end voltage, current, real, reactive, and apparent power, and power factor; (3) percent voltage regulation; and (4) theoretical maximum real power delivered to the receiving end with rated terminal voltages.

The following sample run is for a three-phase 345 kV, 200 km, 60 Hz line with positive sequence impedance 0.047 + j 0.37 Ω/km and positive sequence shunt admittance 0 + j 4.1E-6 S/km. Full-load at the receiving-end of the line is 800 MVA at 0.99 power factor leading and at 328 kV. 75% total shunt reactive compensation is installed, half (37.5%) at each end of the line. The shunt reactors are removed at heavy loads. There are no intermediate substations and there is no series compensation.

SAMPLE RUN   CHAP6   345 kV   200 km   LINE

```
LINE DATA

ENTER THE NAME OF THE LINE(ANY ALPHANUMERIC SEQUENCE) SOFTWARE MANUAL
SAMPLE
IS THIS A THREE PHASE LINE(ENTER Y)OR A SINGLE PHASE LINE(ENTER N) ? y
ENTER THE RATED LINE VOLTAGE(LINE-LINE kV) 345
ENTER THE LINE LENGTH(km) 200
ENTER THE SERIES RESISTANCE(ohms/km) 0.047
ENTER THE SERIES REACTANCE(ohms/km) 0.37
ENTER THE SHUNT CONDUCTANCE(mhos/km) 0
ENTER THE SHUNT SUSCEPTANCE(mhos/km) 4.1e-6


RECEIVING END DATA

ENTER THE RECEIVING END FULL-LOAD VOLTAGE(LINE-LINE kV) 328
ENTER THE RECEIVING END FULL-LOAD APPARENT POWER(MVA) 800
ENTER THE RECEIVING END FULL-LOAD POWER FACTOR(PER UNIT) 0.99
IS THE POWER FACTOR LAGGING(ENTER Y)OR LEADING(ENTER N) ? n


IS THE LINE COMPENSATED(Y OR N) ? y

COMPENSATION INPUT DATA AT EACH LOCATION
IS GIVEN IN PER CENT OF TOTAL SHUNT SUSCEPTANCE
OR SERIES REACTANCE OF THE LINE


ENTER THE % SHUNT REACTIVE COMPENSATION AT THE SENDING END 37.5
ENTER THE % SERIES CAPACITIVE COMPENSATION AT THE SENDING END 0
ENTER THE % SHUNT REACTIVE COMPENSATION AT THE RECEIVING END 37.5
ENTER THE % SERIES CAPACITIVE COMPENSATION AT THE RECEIVING END 0
ENTER THE NUMBER OF INTERMEDIATE SUBSTATIONS(0,1,OR 2): 0

DO YOU WANT TO CONTINUE (Y) OR RESET THE INPUT DATA (N) ? y
```

USE THE PRT SC OPTION NOW IF YOU WANT TO PRINT THE RESULTS
PRESS RETURN TO CONTINUE
************************INPUT DATA**************************
SOFTWARE MANUAL SAMPLE   THREE-PHASE   345 kV   200 km
    LINE DATA:
SERIES RESISTANCE (ohms/km)   4.700E-02
SERIES REACTANCE (ohms/km)   3.700E-01
SHUNT CONDUCTANCE (mhos/km)   0.000E+00
SHUNT SUSCEPTANCE (mhos/km)   4.100E-06

RECEIVING END DATA:
RECEIVING END FULL-LOAD VOLTAGE (kV)   3.280E+02
RECEIVING END FULL-LOAD
APPARENT POWER (MVA)   8.000E+02
RECEIVING END FULL-LOAD
POWER FACTOR (per unit)   9.900E-01LEADING

COMPENSATION DATA:
SHUNT COMPENSATION AT SENDING END(%)   37.5
SERIES COMPENSATION AT SENDING END(%)   0
SHUNT COMPENSATION AT RECEIVING END(%)   37.5
SERIES COMPENSATION AT RECEIVING END(%)   0
NUMBER OF INTERMEDIATE SUBSTATIONS   0
TOTAL SHUNT COMPENSATION(%)   75
TOTAL SERIES COMPENSATION(%)   0

```
**********************OUTPUT DATA**************************

   *****WITHOUT COMPENSATION*****
CHARACTERISTIC IMPEDANCE(ohms)              3.016E+02/ -3.620E+00DEG
PROPAGATION CONSTANT(1/m)                   7.807E-05 + j 1.234E-03
WAVELENGTH(km)                              5.091E+03
SURGE IMPEDANCE LOADING(MW)                 3.946E+02


EQUIVALENT PI CIRCUIT:
SERIES IMPEDANCE(ohms)                      9.211E+00   +j 7.327E+01
SHUNT ADMITTANCE(mhos)                      5.336E-07   +j 8.242E-04
ABCD PARAMETERS:
A PARAMETER(per unit)                       9.698E-01/  2.254E-01DEG
B PARAMETER(ohms)                           7.384E+01/  8.283E+01DEG
C PARAMETER(mhos)                           8.117E-04/  9.007E+01DEG
D PARAMETER(per unit)                       9.698E-01/  2.254E-01DEG

      ******WITH COMPENSATION******
FULL-LOAD ABCD PARAMETERS
(SHUNT COMPENSATION IS REMOVED):
A PARAMETER (per unit)                      9.698E-01/  2.254E-01DEG
B PARAMETER (ohms)                          7.384E+01/  8.283E+01DEG
C PARAMETER(mhos)                           8.117E-04/  9.007E+01DEG
D PARAMETER (per unit)                      9.698E-01/  2.254E-01DEG


NO-LOAD ABCD PARAMETERS
(SHUNT AND SERIES COMPENSATION ARE INCLUDED):
A PARAMETER (per unit)                      9.925E-01/  5.591E-02DEG
B PARAMETER(ohms)                           7.384E+01/  8.283E+01DEG
C PARAMETER(mhos)                           2.053E-04/  8.988E+01DEG
D PARAMETR (per unit)                       9.925E-01/  5.591E-02DEG


SENDING END VOLTAGE(LINE-TO-LINE kV)        3.636E+02/  2.992E+01 (+30 ) DEG
SENDING END CURRENT(kA)                     1.396E+00/  1.459E+01DEG
SENDING END REAL POWER(MW)                  8.479E+02
SENDING END REACTIVE POWER(Mvars)           2.324E+02
SENDING END APPARENT POWER(MVA)             8.792E+02
SENDING END POWER FACTOR(per unit)          9.644E-01 LAGGING
VOLTAGE REGULATION(%)                       1.169E+01
THEORETICAL MAXIMUM REAL POWER
DELIVERED(MW)                               1.411E+03


DO YOU WANT TO RUN IT AGAIN(ENTER Y)OR QUIT(ENTER N)? n
```

6. POWER FLOW (CHAP7)

This program computes the voltage magnitude and phase angle at each bus in a power system under balanced three-phase steady-state operation. Bus voltages and phase angles are then used to compute generator, line, and transformer loadings.

The maximum number of buses is 100. A 30-bus system can be run on a computer with 256K of RAM. More RAM (up to 1 Meg for 100 buses) is required for more than 30 buses. Also, the maximum number of transmission lines or transformers is 100.

Input data include bus data, transmission line data, and transformer data, which are stored in three data files. The user can store the input/output data for up to three separate cases. The bus, line, and transformer input data files for Case 1 are called PFB1.DAT, PFL1.DAT, and PFT1.DAT. Input data files for Case 2 (3) have similar names, except that the number 2 (3) is used.

Bus input data include: (1) bus number, (2) bus type (0 for the swing bus, 1 for a load bus, 2 for a voltage-controlled bus, and 3 for a voltage-controlled bus with a tap-changing transformer), (3) per-unit voltage magnitude for the swing bus and for voltage-controlled buses, (4) phase angle (degrees) for the swing bus, (5) per-unit real generation, (6) per-unit reactive generation; (7) per-unit real load, (8) per-unit reactive load, and (9) maximum and minimum limits on per-unit reactive generation for voltage-controlled (type 2) buses. The swing bus is always bus1 for this program.

Line input data include: (1) line number, (2) the two buses to which the line is connected, (3) per-unit series resistance, (4) per-unit series reactance, (5) per-unit shunt conductance, (6) per-unit shunt susceptance, and (7) per-unit maximum line loading (apparent power).

Transformer data include: (1) transformer number, (2) the two buses to which the transformer is connected, (3) per-unit winding resistance, (4) per-unit leakage reactance, (5) per-unit core-loss conductance, (6) per-unit magnetizing susceptance, (7) per-unit maximum transformer loading (apparent power), and (8) per-unit maximum tap setting (1.0 to 1.25).

The user has the following options for each input data file: (1) initialize the file (which removes all data currently stored); (2) enter or change the data for a selected bus, line, or transformer; (3) remove a bus, line, or transformer; (4) display the data for one bus, line, or transformer; and (5) display the data for all buses, lines, or transformers starting with a given one.

Type 3 Bus

A voltage-controlled bus with a tap-changing transformer is designated when the user selects bus type 3 in the bus input data. For type 3 buses, there must be a transformer in the transformer input data file whose second bus number is the voltage-controlled bus. The program will warn the user when there is a type 3 bus and no transformer whose second bus number is the type 3 bus. Also, the program assumes that the tap is on the same side of the transformer as the second bus number of that transformer.

The program varies the tap setting of each tap-changing transformer connected to a type 3 bus until either: (1) the computed bus voltage is within ±0.4% of the input bus voltage, or (2) the maximum or minimum tap setting is reached. The user inputs a value from 1.0 to 1.25 for the maximum tap setting MAXTAP of each transformer in the transformer input data file. The program sets the minimum tap setting to MINTAP = (2.0 - MAXTAP).

Fixed Transformer Tap Settings

The user can also set tap settings of tap-changing transformers to fixed values. The program asks the user if any of the transformers are tap-changing transformers with fixed taps. If a "Y" (yes) answer is given, the user selects the fixed-tap transformers and then inputs the tap settings (values from 0.75 to 1.25). The program assumes that the tap is on the same side of the transformer as the <u>second</u> bus number of that transformer.

The program uses the Newton-Raphson iterative method for solving the power-flow. The user can select the maximum number of iterations (default value is 20) and the power mismatch for convergence (default value is 0.0001 per unit).

Starting Values

The user has the following options for starting values: (1) flat start, or (2) output values from the previous run. If the user selects (1), then the initial voltage magnitudes of type 1 buses are set equal to the swing bus voltage, and all initial phase angles are set equal to the swing bus phase angle. Also, the initial tap settings of all variable tap-changing transformers are set equal to 1.0. If the user selects (2), then the initial bus voltage magnitudes and phase angles as well as initial transformer tap settings are set equal to the values from the bus output and transformer output data files.

When preparing a "base case" for the first time with no output data, then select (1), a flat start. After running the program for a base case with a successful power-flow solution stored in the output data files, select (2) when you make a change such as removing a line or transformer. The program will usually converge more rapidly when you start with the output values from a previous (and successful) run.

Output data include bus output, line output, and transformer output, which are stored in three separate data files. For Case 1, the output data files are called PFBO1.DAT, PFLO1.DAT, and PFTO1.DAT. Similar names are used for Case 2 (3) except the number 2 (3) is used. The program gives the user the option of displaying the data on the monitor and on the printer.

Bus output data include: (1) bus number, (2) per-unit voltage magnitude, (3) voltage phase angle (degrees), (4) per-unit real generation, (5) per-unit reactive generation, (6) per-unit real load power, and (7) per-unit reactive load power. Those buses for which the voltage magnitude is 5% higher or lower than the swing bus voltage are also identified.

Line output data include: (1) line number, (2) the two buses to which the line is connected, (3) the per-unit real, reactive, and apparent power flows into each end of the line. Those lines whose loadings exceed their maximum loadings are also identified.

Transformer output data include: (1) transformer number, (2) the two buses to which the transformer is connected, (3) the per-unit real, reactive, and apparent power flows into each winding, and (4) per-unit tap setting. Those transformers whose loadings exceed their maximum loadings are also identified.

The following sample run is for the five-bus power system whose single-line diagram is shown in Figure 6a. Bus, line, and transformer input data are given in the sample run. Note that bus 5 is a type 3 bus. The program adjusts the tap setting on transformer T1.

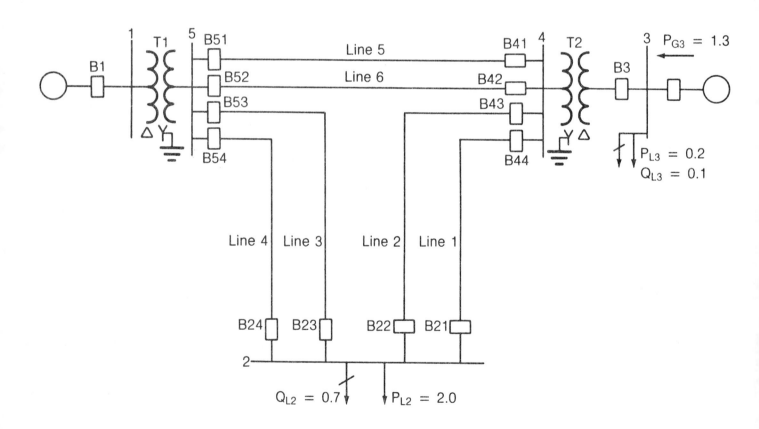

Figure 6a Single-line diagram for SAMPLE RUN CHAP7

(per-unit values are shown)

WHAT IS YOUR CASE NUMBER (ENTER AN INTEGER FROM 1 TO 3) ? 1

ENTER YOUR CASE NAME (ANY ALPHANUMERIC SEQUENCE) SAMPLE RUN

WOULD YOU LIKE TO:
        1. UPDATE THE INPUT DATA FILES
        2. RUN THE PROGRAM
        3. DISPLAY THE OUTPUT DATA
        4. STOP

ENTER YOUR SELECTION (1,2,3 OR 4) 1

        SELECTIONS:
            1. BUS DATA
            2. LINE DATA
            3. TRANSFORMER DATA
            4. DONE WITH DATA

ENTER YOUR SELECTION (1,2,3 OR 4): 1
SELECTION:

    1. INITIALIZE THE BUS DATA FILE
    2. ADD A NEW BUS OR CHANGE AN EXISTING BUS
    3. REMOVE A BUS
    4. DISPLAY ONE BUS
    5. DISPLAY ALL BUSES BELOW A GIVEN BUS
    6. DONE WITH BUS DATA

 ENTER YOUR SELECTION : 5
 ENTER THE NUMBER OF THE FIRST BUS TO BE DISPLAYED : 1

USE THE Ctrl PRINT SCREEN OPTION NOW IF YOU WANT TO PRINT THE RESULTS.

PRESS RETURN TO CONTINUE

BUS INPUT DATA FOR   SAMPLE RUN

BUS#	TYPE	V	DELTA	PG	QG	PL	QL	QGMAX	QGMIN
		per unit	degrees	per unit	per unit	per unit	per unit	per unit	per unit
1	0	1.000	0.000	---	---	0.000	0.000	---	---
2	1	---	---	0.000	0.000	2.000	0.200	---	---
3	2	1.040	---	1.300	---	0.200	0.100	1.000	-0.700
4	1	---	---	0.000	0.000	0.000	0.000	---	---
5	3	1.020	---	0.000	0.000	0.000	0.000	---	---

REMOVE Ctrl PRINT SCREEN AND THEN PRESS RETURN TO CONTINUE
SELECTION:

    1. INITIALIZE THE BUS DATA FILE
    2. ADD A NEW BUS OR CHANGE AN EXISTING BUS
    3. REMOVE A BUS
    4. DISPLAY ONE BUS
    5. DISPLAY ALL BUSES BELOW A GIVEN BUS
    6. DONE WITH BUS DATA

ENTER YOUR SELECTION : 6

    SELECTIONS:
        1. BUS DATA
        2. LINE DATA
        3. TRANSFORMER DATA
        4. DONE WITH DATA

ENTER YOUR SELECTION (1,2,3 OR 4): 2
SELECTIONS:

    1. INITIALIZE THE LINE DATA FILE
    2. ADD A NEW LINE OR CHANGE AN EXISTING LINE
    3. REMOVE A LINE
    4. DISPLAY ONE LINE
    5. DISPLAY ALL LINES BELOW A GIVEN LINE
    6. DONE WITH LINE DATA

ENTER YOUR SELECTION: 5
ENTER THE NUMBER OF THE FIRST LINE TO BE DISPLAYED: 1

USE THE Ctrl PRINT SCREEN OPTION NOW IF YOU WANT TO PRINT THE RESULTS.

PRESS RETURN TO CONTINUE

LINE INPUT DATA FOR   SAMPLE RUN

LINE#	BUS-TO-BUS		R	X	G	B	MAXMVA
			per unit	per unit	per unit	per unit	per unit
1	2	4	0.036	0.400	0.000	0.430	3.000
2	2	4	0.036	0.400	0.000	0.430	3.000
3	2	5	0.018	0.200	0.000	0.220	3.000
4	2	5	0.018	0.200	0.000	0.220	3.000
5	4	5	0.009	0.100	0.000	0.110	3.000
6	4	5	0.009	0.100	0.000	0.110	3.000

REMOVE Ctrl PRINT SCREEN AND THEN PRESS RETURN TO CONTINUE
SELECTIONS:

    1. INITIALIZE THE LINE DATA FILE
    2. ADD A NEW LINE OR CHANGE AN EXISTING LINE
    3. REMOVE A LINE
    4. DISPLAY ONE LINE
    5. DISPLAY ALL LINES BELOW A GIVEN LINE
    6. DONE WITH LINE DATA

ENTER YOUR SELECTION: 6

        SELECTIONS:
            1. BUS DATA
            2. LINE DATA
            3. TRANSFORMER DATA
            4. DONE WITH DATA

ENTER YOUR SELECTION (1,2,3 OR 4): 3
 SELECTIONS:

    1. INITIALIZE THE TRANSFORMER DATA FILE
    2. ADD A NEW TRANSFORMER OR CHANGE AN EXISTING TRANSFORMER
    3. REMOVE A TRANSFORMER
    4. DISPLAY ONE TRANSFORMER
    5. DISPLAY ALL TRANSFORMERS BELOW A GIVEN TRANSFORMER
    6. DONE WITH TRANSFORMER DATA

 ENTER YOUR SELECTION : 5
 ENTER THE NUMBER OF THE FIRST TRANSFORMER TO BE DISPLAYED :1

USE THE Ctrl PRINT SCREEN OPTION NOW IF YOU WANT TO PRINT THE RESULTS.

PRESS RETURN TO CONTINUE
                    TRANSFORMER INPUT DATA FOR SAMPLE RUN

TRAN.#	BUS-TO-BUS		R	X	G	B	MAXMVA	MAXTAP
			per unit	per unit	per unit	per unit	per unit	per unit
1	1	5	0.006	0.080	0.000	0.000	1.500	1.200
2	3	4	0.003	0.040	0.000	0.000	2.500	1.000

REMOVE Ctrl PRINT SCREEN AND THEN PRESS RETURN TO CONTINUE
SELECTIONS:

        1. INITIALIZE THE TRANSFORMER DATA FILE
        2. ADD A NEW TRANSFORMER OR CHANGE AN EXISTING TRANSFORMER
        3. REMOVE A TRANSFORMER
        4. DISPLAY ONE TRANSFORMER
        5. DISPLAY ALL TRANSFORMERS BELOW A GIVEN TRANSFORMER
        6. DONE WITH TRANSFORMER DATA

ENTER YOUR SELECTION : 6

        SELECTIONS:
                1. BUS DATA
                2. LINE DATA
                3. TRANSFORMER DATA
                4. DONE WITH DATA

ENTER YOUR SELECTION (1,2,3 OR 4): 4

WOULD YOU LIKE TO:
        1. UPDATE THE INPUT DATA FILES
        2. RUN THE PROGRAM
        3. DISPLAY THE OUTPUT DATA
        4. STOP

ENTER YOUR SELECTION (1,2,3 OR 4) 2

DO ANY OF THE TRANSFORMERS HAVE TAP-CHANGERS WITH FIXED TAP SETTINGS

(Y OR N) ? N

  THE TOLERANCE LEVEL IS   .0001 .

  DO YOU WANT TO CHANGE THE TOLERANCE LEVEL (Y OR N)? N

  THE MAXIMUM NUMBER OF ITERATIONS IS   20 .

  DO YOU WANT TO CHANGE THE NUMBER OF ITERATIONS (Y OR N)? N
SELECTION OF INITIAL BUS VOLTAGE MAGNITUDES AND ANGLES :

    1. FLAT START
    2. OUTPUT VALUES FROM THE PREVIOUS RUN

ENTER YOUR SELECTION (1 OR 2) 1

```
***STARTING VALUES***
 1   0
 1   0
 1.04   0
 1   0
 1.02   0

 AT( 1 ) =  1
 AT( 2 ) =  1
****THE PROGRAM IS RUNNING****


TOTAL NUMBER OF ITERATIONS =   14

MISMATCH =   9.23E-06

PRESS RETURN TO CONTINUE


**THE RESULTS ARE NOW BEING SENT TO OUTPUT DATA FILES.**




WOULD YOU LIKE TO:
      1. UPDATE THE INPUT DATA FILES
      2. RUN THE PROGRAM
      3. DISPLAY THE OUTPUT DATA
      4. STOP



ENTER YOUR SELECTION (1,2,3 OR 4) 3



DO YOU WANT TO DISPLAY THE BUS OUTPUT DATA ( Y OR N ) ? Y

DO YOU WANT TO DISPLAY THE LINE OUTPUT DATA ( Y OR N ) ? Y

DO YOU WANT TO DISPLAY THE TRANSFORMER OUTPUT DATA ( Y OR N ) ? Y

DO YOU WANT TO DISPLAY THE OUTPUT DATA IN EXPONENTIAL FORMAT

( Y OR N ) ? N

USE THE Ctrl PRINT SCREEN OPTION NOW IF YOU WANT TO PRINT THE RESULTS.
PRESS RETURN TO CONTINUE.
```

POWER FLOW BUS OUTPUT DATA FOR SAMPLE RUN

			GENERATION		LOAD	
BUS#	VOLTAGE MAGNITUDE	PHASE ANGLE	PG	QG	PL	QL .95>V>1.05
	per unit	degrees	per unit	per unit	per unit	per unit
1	1.000	0.000	0.938	-0.844	0.000	0.000
2	1.042	-11.333	0.000	0.000	2.000	0.200
3	1.040	-1.116	1.300	-0.044	0.200	0.100
4	1.043	-3.463	0.000	0.000	0.000	0.000
5	1.022	-4.314	0.000	0.000	0.000	0.000
		TOTAL	2.238	-0.888	2.200	0.300

POWER FLOW LINE OUTPUT DATA FOR SAMPLE RUN

LINE #	BUS	TO BUS	P	Q	S	RATING EXCEEDED
1	2	4	-0.367	-0.179	0.408	
	4	2	0.372	-0.238	0.441	
2	2	4	-0.367	-0.179	0.408	
	4	2	0.372	-0.238	0.441	
3	2	5	-0.633	0.079	0.638	
	5	2	0.640	-0.232	0.681	
4	2	5	-0.633	0.079	0.638	
	5	2	0.640	-0.232	0.681	
5	4	5	0.177	0.143	0.227	
	5	4	-0.176	-0.254	0.309	
6	4	5	0.177	0.143	0.227	
	5	4	-0.176	-0.254	0.309	

POWER FLOW TRANSFORMER OUTPUT DATA FOR SAMPLE RUN

TRAN.#	BUS	TO BUS	P	Q	S	TAP SETTING	RATING EXCEEDED
1	1	5	0.977	-0.326	1.030	0.960	
	5	1	-0.971	0.408	1.053		
2	3	4	1.100	-0.144	1.109	1.000	
	4	3	-1.097	0.189	1.113		

REMOVE Ctrl PRINT SCREEN AND THEN PRESS RETURN TO CONTINUE.

WOULD YOU LIKE TO:
    1. UPDATE THE INPUT DATA FILES
    2. RUN THE PROGRAM
    3. DISPLAY THE OUTPUT DATA
    4. STOP

ENTER YOUR SELECTION (1,2,3 OR 4) 4

7. SYMMETRICAL SHORT CIRCUITS (CHAP8)

This program computes the ac or symmetrical fault current for a bolted three-phase short circuit at any bus in an N-bus power system. The program also computes contributions to the fault current from synchronous machines, transmission lines, and transformers connected to the faulted bus. It also computes the bus voltages during the fault as well as the positive-sequence bus impedance matrix.

The maximum number of buses is 100. A 50-bus system can be run on a computer with 256K of RAM. More RAM (up to 500K for100 buses) is required for more than 50 buses. Also, the maximum number of lines or transformers is 100.

Input data include synchronous machine data, transmission line data, and transformer data, which are stored in three data files. The user can store the input data for up to five separate cases. The machine, line, and transformer input data files for Case 1 are called SSCM1.DAT, SSCL1.DAT, and SSCT1.DAT. Input data files for Case 2 (3,4,5) have similar names, except the number 2 (3,4,5) is used.

Synchronous machine input data include the machine number, bus number, and per-unit positive-sequence reactance of each machine. When the machine reactances are subtransient reactances, the program computes subtransient fault currents. Alternatively, when the machine reactances are transient or synchronous reactances, the program computes transient or steady-state fault currents.

Transmission line data include the line number, the two buses to which the line is connected and the per-unit positive-sequnce reactance for each line. Similarly, transformer input data include the transformer number, the two buses to which the transformer is connected, and the per-unit leakage reactance for each transformer.

The program user has the following three options: (1) update the input data files, (2) run the program, and (3) stop.

When the user updates the input data files, the following options for each input data file are given: (1) initialize the file (which removes all data currently stored); (2) enter or change the data for a selected machine, line or transformer; (3) remove a machine, line or transformer; (4) display the data for one machine, line or transformer; and (5) display the data for all machines, lines or transformers after a given one.

When the user runs the program, the per-unit prefault voltage V_F is selected. The program then computes the positive-sequence bus impedance matrix using the one-step-at-a-time method. After Z_{bus} is computed, the fault currents, contributions to the fault currents, and bus voltages during the fault are computed for a fault at bus 1, then bus 2, ... up to the final bus N. All computations are in per unit.

The user has the following three output display options: (1) display the fault currents, (2) display the bus voltages during the fault, and (3) display the bus-impedance matrix. The outputs can be displayed in exponential format (scientific notation) or fixed-point format (numbers with decimal points).

The following sample run computes the fault currents for the 5-bus power system whose single-line diagram is shown in Figure 6a. Machine, line, and transformer input data are given in the sample run. The prefault voltage is $V_F = 1.0$ per unit.

SAMPLE RUN 7.1 CHAP8 FIGURE 6a


```
ENTER YOUR CASE NUMBER ( 1,2,3,4 OR 5 ) 1

ENTER YOUR CASE NAME(ANY ALPHANUMERIC SEQUENCE)SAMPLE RUN 7.1
SELECTIONS:

     1. UPDATE THE INPUT DATA
     2. RUN THE PROGRAM
     3. STOP


ENTER YOUR SELECTION ( 1,2 OR 3 ) 1
THESE ARE YOUR INPUT DATA SELECTIONS:

     1. UPDATE THE SYNCHRONOUS MACHINE INPUT DATA
     2. UPDATE THE TRANSMISSION LINE DATA
     3. UPDATE THE TRANSFORMER INPUT DATA
     4. DONE WITH INPUT DATA


ENTER YOUR SELECTION ( 1,2, 3 OR 4 ) 1
CHOICES :

     1. INITIALIZE FILE
     2. ADD A NEW MACHINE OR CHANGE AN EXISTING MACHINE
     3. REMOVE A MACHINE
     4. DISPLAY ONE MACHINE
     5. DISPLAY ALL MACHINES AFTER A GIVEN MACHINE
     6. DONE WITH MACHINE DATA


ENTER YOUR CHOICE ( 1,2,3,4,5 OR 6 ) : 5
ENTER THE NUMBER OF THE FIRST MACHINE TO BE DISPLAYED 1

USE Ctrl PRINT SCREEN OPTION IF YOU WANT TO PRINT THE RESULTS.

PRESS RETURN TO CONTINUE.
```

MACHINE#	BUS	X1
		per unit
1	1	0.1800
2	3	0.0900

REMOVE Ctrl PRINT SCREEN AND THEN PRESS RETURN TO CONTINUE
CHOICES :

    1. INITIALIZE FILE
    2. ADD A NEW MACHINE OR CHANGE AN EXISTING MACHINE
    3. REMOVE A MACHINE
    4. DISPLAY ONE MACHINE
    5. DISPLAY ALL MACHINES AFTER A GIVEN MACHINE
    6. DONE WITH MACHINE DATA

ENTER YOUR CHOICE (1,2,3,4,5 OR 6) : 6
THESE ARE YOUR INPUT DATA SELECTIONS:

    1. UPDATE THE SYNCHRONOUS MACHINE INPUT DATA
    2. UPDATE THE TRANSMISSION LINE DATA
    3. UPDATE THE TRANSFORMER INPUT DATA
    4. DONE WITH INPUT DATA

ENTER YOUR SELECTION (1,2, 3 OR 4) 2
CHOICES :

    1. INITIALIZE FILE
    2. ADD A NEW LINE OR CHANGE AN EXISTING LINE
    3. REMOVE A LINE
    4. DISPLAY ONE LINE
    5. DISPLAY ALL LINES AFTER A GIVEN LINE
    6. DONE WITH TRANSMISSION LINE DATA

ENTER YOUR CHOICE (1,2,3,4,5 OR 6) : 5
ENTER THE NUMBER OF THE FIRST LINE TO BE DISPLAYED 1

USE THE Ctrl PRINT SCREEN OPTION TO PRINT THE RESULTS.

PRESS RETURN TO CONTINUE.

TRANSMISSION LINE INPUT DATA FOR SAMPLE RUN 7.1

LINE#	BUS	-TO-	BUS	X1
				per unit
1	2		4	0.4000
2	2		4	0.4000
3	2		5	0.2000
4	2		5	0.2000
5	4		5	0.1000
6	4		5	0.1000

REMOVE Ctrl PRINT SCREEN AND THEN PRESS RETURN TO CONTINUE
CHOICES :

    1. INITIALIZE FILE
    2. ADD A NEW LINE OR CHANGE AN EXISTING LINE
    3. REMOVE A LINE
    4. DISPLAY ONE LINE
    5. DISPLAY ALL LINES AFTER A GIVEN LINE
    6. DONE WITH TRANSMISSION LINE DATA

ENTER YOUR CHOICE (1,2,3,4,5 OR 6) : 6
THESE ARE YOUR INPUT DATA SELECTIONS:

    1. UPDATE THE SYNCHRONOUS MACHINE INPUT DATA
    2. UPDATE THE TRANSMISSION LINE DATA
    3. UPDATE THE TRANSFORMER INPUT DATA
    4. DONE WITH INPUT DATA

ENTER YOUR SELECTION (1,2, 3 OR 4) 3
CHOICES :

    1. INITIALIZE FILE
    2. ADD A NEW TRANSFORMER OR CHANGE AN EXISTING TRANSFORMER
    3. REMOVE A TRANSFORMER
    4. DISPLAY ONE TRANSFORMER
    5. DISPLAY ALL TRANSFORMERS AFTER A GIVEN LINE
    6. DONE WITH TRANSFORMER DATA

ENTER YOUR CHOICE (1,2,3,4,5 OR 6) : 5
ENTER THE NUMBER OF THE FIRST TRANSFORMER TO BE DISPLAYED 1

USE THE Ctrl PRINT SCREEN OPTION IF YOU WANT TO PRINT THE RESULTS.

PRESS RETURN TO CONTINUE.

TRANSFORMER INPUT DATA FOR SAMPLE RUN 7.1

TRANSF#	BUS-TO-BUS		XL
			per unit
1	1	5	0.0800
2	3	4	0.0400

REMOVE Ctrl PRINT SCREEN AND THEN PRESS RETURN TO CONTINUE
CHOICES :

    1. INITIALIZE FILE
    2. ADD A NEW TRANSFORMER OR CHANGE AN EXISTING TRANSFORMER
    3. REMOVE A TRANSFORMER
    4. DISPLAY ONE TRANSFORMER
    5. DISPLAY ALL TRANSFORMERS AFTER A GIVEN LINE
    6. DONE WITH TRANSFORMER DATA

ENTER YOUR CHOICE (1,2,3,4,5 OR 6) : 6
THESE ARE YOUR INPUT DATA SELECTIONS:

    1. UPDATE THE SYNCHRONOUS MACHINE INPUT DATA
    2. UPDATE THE TRANSMISSION LINE DATA
    3. UPDATE THE TRANSFORMER INPUT DATA
    4. DONE WITH INPUT DATA

ENTER YOUR SELECTION (1,2, 3 OR 4) 4
SELECTIONS:

    1. UPDATE THE INPUT DATA
    2. RUN THE PROGRAM
    3. STOP

ENTER YOUR SELECTION (1,2 OR 3) 2

WHAT IS THE PREFAULT VOLTAGE (per unit)? 1.0

DISPLAY THE FAULT CURRENTS (Y OR N)? Y

DISPLAY THE BUS VOLTAGES DURING THE FAULT (Y OR N)? Y

DISPLAY THE BUS IMPEDANCE MATRICES (Y OR N)? Y

DO YOU WANT THE OUTPUTS PRINTED IN EXPONENTIAL FORMAT (Y OR N) ? N

DO YOU WANT TO CONTINUE (Y) OR RESET THE DATA (N) ? Y

USE THE Ctrl PRINT SCREEN OPTION NOW IF YOU WANT TO PRINT THE RESULTS.
PRESS RETURN TO CONTINUE

FAULT CURRENTS FOR SAMPLE RUN 7.1

FAULT BUS	THREE-PHASE FAULT CURRENT	GEN LINE OR TRSF	BUS-TO-BUS		CONTRIBUTIONS TO FAULT CURRENT
	per unit				per unit
1	9.510				
		G 1	GRND –	1	5.5556
		T 1	5 –	1	3.9548
2	6.344				
		L 1	4 –	2	1.1935
		L 2	4 –	2	1.1935
		L 3	5 –	2	1.9786
		L 4	5 –	2	1.9786
3	14.028				
		G 2	GRND –	3	11.1111
		T 2	4 –	3	2.9167
4	10.994				
		L 1	2 –	4	0.2358
		L 2	2 –	4	0.2358
		L 5	5 –	4	1.4151
		L 6	5 –	4	1.4151
		T 2	3 –	4	7.6923
5	9.631				
		L 3	2 –	5	0.4132
		L 4	2 –	5	0.4132
		L 5	4 –	5	2.4793
		L 6	4 –	5	2.4793
		T 1	1 –	5	3.8462

VF =   1   PER-UNIT BUS VOLTAGES DURING FAULTS

SAMPLE RUN 7.1

FAULT
BUS   BUS1(6..) BUS2(7..) BUS3(8..) BUS4(9..) BUS5(10..)

1   0.0000   0.3729   0.6441   0.4859   0.3164
2   0.5817   0.0000   0.6382   0.4774   0.3957
3   0.4750   0.2000   0.0000   0.1167   0.2417
4   0.4057   0.0943   0.3077   0.0000   0.1415
5   0.3077   0.0826   0.4793   0.2479   0.0000

PER-UNIT ZBUS FOR SAMPLE RUN 8.1

ROW 1   0.10515   0.06594   0.03743   0.05406   0.07188
ROW 2   0.06594   0.15762   0.05703   0.08238   0.09525
ROW 3   0.03743   0.05703   0.07129   0.06297   0.05406
ROW 4   0.05406   0.08238   0.06297   0.09096   0.07809
ROW 5   0.07188   0.09525   0.05406   0.07809   0.10383

REMOVE THE Ctrl PRINT SCREEN OPTION AND PRESS RETURN TO CONTINUE
SELECTIONS:

    1. UPDATE THE INPUT DATA
    2. RUN THE PROGRAM
    3. STOP

ENTER YOUR SELECTION (1,2 OR 3) 3
Ok

The following sample run computes the bus impedance matrix for the 3-bus circuit shown in Figure 7a. The Z_{bus} matrix displayed in the output is the same as that given in Problem 8.8 of the text *Power System Analysis and Design With Personal Computer Applications* , by J.D. Glover and M. Sarma. After computing Z_{bus} from this sample run, the user can then work Problem 8.16 in the text.

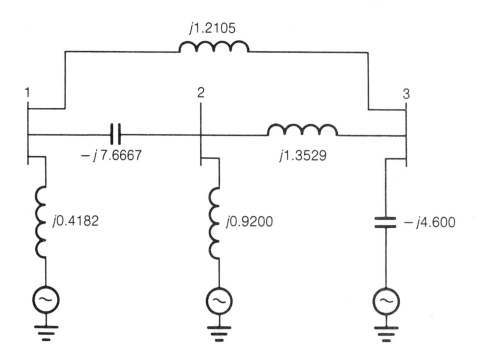

Figure 7a Circuit diagram for SAMPLE RUN 7.2 : CHAP8

(per-unit reactances are shown)

SAMPLE RUN 7.2   CHAP8   FIGURE 7a

ENTER YOUR CASE NUMBER (1,2,3,4 OR 5) 2

ENTER YOUR CASE NAME(ANY ALPHANUMERIC SEQUENCE)SAMPLE RUN 7.2
SELECTIONS:

    1. UPDATE THE INPUT DATA
    2. RUN THE PROGRAM
    3. STOP

ENTER YOUR SELECTION (1,2 OR 3) 1
THESE ARE YOUR INPUT DATA SELECTIONS:

    1. UPDATE THE SYNCHRONOUS MACHINE INPUT DATA
    2. UPDATE THE TRANSMISSION LINE DATA
    3. UPDATE THE TRANSFORMER INPUT DATA
    4. DONE WITH INPUT DATA

ENTER YOUR SELECTION (1,2, 3 OR 4) 1
CHOICES :

    1. INITIALIZE FILE
    2. ADD A NEW MACHINE OR CHANGE AN EXISTING MACHINE
    3. REMOVE A MACHINE
    4. DISPLAY ONE MACHINE
    5. DISPLAY ALL MACHINES AFTER A GIVEN MACHINE
    6. DONE WITH MACHINE DATA

ENTER YOUR CHOICE (1,2,3,4,5 OR 6) : 5
ENTER THE NUMBER OF THE FIRST MACHINE TO BE DISPLAYED 1

USE Ctrl PRINT SCREEN OPTION IF YOU WANT TO PRINT THE RESULTS.

PRESS RETURN TO CONTINUE.

MACHINE#	BUS	X1
		per unit
1	1	0.4182
2	2	0.9200
3	3	-4.6000

REMOVE Ctrl PRINT SCREEN AND THEN PRESS RETURN TO CONTINUE
CHOICES :

    1. INITIALIZE FILE
    2. ADD A NEW MACHINE OR CHANGE AN EXISTING MACHINE
    3. REMOVE A MACHINE
    4. DISPLAY ONE MACHINE
    5. DISPLAY ALL MACHINES AFTER A GIVEN MACHINE
    6. DONE WITH MACHINE DATA

ENTER YOUR CHOICE (1,2,3,4,5 OR 6) : 6
THESE ARE YOUR INPUT DATA SELECTIONS:

    1. UPDATE THE SYNCHRONOUS MACHINE INPUT DATA
    2. UPDATE THE TRANSMISSION LINE DATA
    3. UPDATE THE TRANSFORMER INPUT DATA
    4. DONE WITH INPUT DATA

ENTER YOUR SELECTION (1,2, 3 OR 4) 2
CHOICES :

    1. INITIALIZE FILE
    2. ADD A NEW LINE OR CHANGE AN EXISTING LINE
    3. REMOVE A LINE
    4. DISPLAY ONE LINE
    5. DISPLAY ALL LINES AFTER A GIVEN LINE
    6. DONE WITH TRANSMISSION LINE DATA

ENTER YOUR CHOICE (1,2,3,4,5 OR 6) : 5
ENTER THE NUMBER OF THE FIRST LINE TO BE DISPLAYED 1

USE THE Ctrl PRINT SCREEN OPTION TO PRINT THE RESULTS.

PRESS RETURN TO CONTINUE.

TRANSMISSION LINE INPUT DATA FOR SAMPLE RUN 7.2

LINE#	BUS	-TO- BUS	X1
			per unit
1	1	2	-7.6667
2	2	3	1.3529
3	1	3	1.2105

REMOVE Ctrl PRINT SCREEN AND THEN PRESS RETURN TO CONTINUE
CHOICES :

    1. INITIALIZE FILE
    2. ADD A NEW LINE OR CHANGE AN EXISTING LINE
    3. REMOVE A LINE
    4. DISPLAY ONE LINE
    5. DISPLAY ALL LINES AFTER A GIVEN LINE
    6. DONE WITH TRANSMISSION LINE DATA

ENTER YOUR CHOICE (1,2,3,4,5 OR 6) : 6
THESE ARE YOUR INPUT DATA SELECTIONS:

    1. UPDATE THE SYNCHRONOUS MACHINE INPUT DATA
    2. UPDATE THE TRANSMISSION LINE DATA
    3. UPDATE THE TRANSFORMER INPUT DATA
    4. DONE WITH INPUT DATA

ENTER YOUR SELECTION (1,2, 3 OR 4) 4
SELECTIONS:

    1. UPDATE THE INPUT DATA
    2. RUN THE PROGRAM
    3. STOP

ENTER YOUR SELECTION (1,2 OR 3) 2

WHAT IS THE PREFAULT VOLTAGE (per unit)? 1

DISPLAY THE FAULT CURRENTS (Y OR N)? N

DISPLAY THE BUS VOLTAGES DURING THE FAULT (Y OR N)? N

DISPLAY THE BUS IMPEDANCE MATRICES (Y OR N)? Y

DO YOU WANT THE OUTPUTS PRINTED IN EXPONENTIAL FORMAT (Y OR N) ? N

DO YOU WANT TO CONTINUE (Y) OR RESET THE DATA (N) ? Y

USE THE Ctrl PRINT SCREEN OPTION NOW IF YOU WANT TO PRINT THE RESULTS.
PRESS RETURN TO CONTINUE

PER-UNIT ZBUS FOR SAMPLE RUN 7.2

```
ROW 1              0.40000   0.10000   0.30000
ROW 2              0.10000   0.80000   0.50000
ROW 3              0.30000   0.50000   1.20000
```

SELECTIONS:

        1. UPDATE THE INPUT DATA
        2. RUN THE PROGRAM
        3. STOP

ENTER YOUR SELECTION (1,2 OR 3) 3
Ok

8. SHORT CIRCUITS (CHAP9)

This program computes the ac or symmetrical fault current for one of the following faults in an N-bus power system: balanced three-phase fault, single line-to-ground fault through fault impedance Z_F, line-to-line fault through Z_F, or double line-to-ground fault through Z_F. The program also computes the contributions to the fault current from synchronous machines, transmission lines, and transformers connected to the faulted bus. And it computes the three-phase bus voltages during the fault as well as the zero-, positive-, and negative-sequence bus impedance matrices.

The maximum number of buses is 100. A 30-bus system can be run on a computer with 256K of RAM. More RAM (up to 1 Meg for 100 buses) is required for more than 30 buses. Also, the maximum number of machines, lines, or transformers is 100.

Input data include synchronous machine data, transmission line data, and transformer data, which are stored in three data files. The user can store the input data for up to five separate cases. The machine, line, and transformer input data files for case 1 are called SCM1.DAT, SCL1.DAT, and SCT1.DAT. Input data files for Case 2 (3,4,5) have similar names, except the number 2 (3,4,5) is used.

Synchronous machine input data include machine number, bus number, per-unit zero-, positive-, and negative-sequence machine reactances, and per-unit neutral reactance for each machine. When the positive-sequence machine reactances are subtransient reactances, the program computes subtransient fault currents. Alternatively, when the positive-sequence machine reactances are transient or synchronous reactances, the program computes transient or steady-state fault currents.

Transmission line input data include line number, the two buses to which the line is connected, and the per-unit zero- and positive-sequence reactance for each line. Transformer input data include transformer number, low-voltage bus number, low-voltage connection (delta or wye), high-voltage bus number, high-voltage connection, per-unit leakage reactance, and per-unit neutral reactance of each wye-connection for each transformer.

For a solidly grounded neutral, enter 0 (zero) for neutral reactance in the machine or transformer input data. For an approximation to an open neutral, enter a large value such as 1000 for neutral reactance.

The program user has the following three selections: (1) update the input data files, (2) run the program, and (3) stop.

When the user updates the input data files, the following options for each input data file are given: (1) initialize the file (which removes all data currently stored); (2) enter or change the data for a selected machine, line, or transformer; (3) remove a machine, line, or transformer; (4) display the data for one machine, line, or transformer; and (5) display the data for all machines, lines, or transformers after a given one.

When the user runs the program, one of the following four faults is selected: (1) three-phase fault, (2) single line-to-ground fault through Z_F, (3) line-to-line fault through Z_F, or (4) double line-to-ground fault through Z_F. The user also selects the per-unit resistive and reactive components of the fault impedance, and the per-unit prefault voltage V_F.

The program then computes the zero-, positive-, and negative- sequence bus impedance matrices using the one-step-at-a-time method. After Z_{bus0}, Z_{bus1}, and Z_{bus2} are computed, the fault currents, contributions to the fault currents, and bus voltages during the fault are computed for a fault at bus 1, then bus 2, ... up to the final bus N. All computations are in per unit.

The user has three output display options: (1) display the fault currents and contributions to the fault currents; (2) display the three-phase bus voltages during the faults; and (3) display the zero-, positive-, and negative-sequence bus impedance matrices. The outputs can be displayed in exponential format (scientific notation) or fixed-point format (numbers with decimal points).

The following sample run computes the fault currents for the 5-bus power system whose single-line diagram is shown in Figure 6a. Machine, line, and transformer input data are given in the sample run. A double line-to-ground fault with $Z_F = 0$ and $V_F = 1.0$ is selected.

SAMPLE RUN   CHAP9   FIGURE 6a

```
ENTER YOUR CASE NUMBER ( 1,2,3,4 OR 5 ) 1

ENTER YOUR CASE NAME(ANY ALPHANUMERIC SEQUENCE)SAMPLE RUN
SELECTIONS:

     1. UPDATE THE INPUT DATA
     2. RUN THE PROGRAM
     3. STOP


ENTER YOUR SELECTION ( 1,2 OR 3 ) 1
THESE ARE YOUR INPUT DATA SELECTIONS:

     1. UPDATE THE SYNCHRONOUS MACHINE INPUT DATA
     2. UPDATE THE TRANSMISSION LINE DATA
     3. UPDATE THE TRANSFORMER INPUT DATA
     4. DONE WITH INPUT DATA


ENTER YOUR SELECTION ( 1,2, 3 OR 4 ) 1
CHOICES :

     1. INITIALIZE FILE
     2. ADD A NEW MACHINE OR CHANGE AN EXISTING MACHINE
     3. REMOVE A MACHINE
     4. DISPLAY ONE MACHINE
     5. DISPLAY ALL MACHINES AFTER A GIVEN MACHINE
     6. DONE WITH MACHINE DATA


ENTER YOUR CHOICE ( 1,2,3,4,5 OR 6 ) : 5
ENTER THE NUMBER OF THE FIRST MACHINE TO BE DISPLAYED 1

USE Ctrl PRINT SCREEN OPTION IF YOU WANT TO PRINT THE RESULTS.

PRESS RETURN TO CONTINUE.
```

SYNCHRONOUS MACHINE INPUT DATA FOR SAMPLE RUN

MACHINE#	BUS	X0	X1	X2	XN
		per unit	per unit	per unit	per unit
1	1	0.0500	0.1800	0.1800	0.0000
2	3	0.0200	0.0900	0.0900	0.0100

REMOVE Ctrl PRINT SCREEN AND THEN PRESS RETURN TO CONTINUE
CHOICES :

    1. INITIALIZE FILE
    2. ADD A NEW MACHINE OR CHANGE AN EXISTING MACHINE
    3. REMOVE A MACHINE
    4. DISPLAY ONE MACHINE
    5. DISPLAY ALL MACHINES AFTER A GIVEN MACHINE
    6. DONE WITH MACHINE DATA

ENTER YOUR CHOICE (1,2,3,4,5 OR 6) : 6
THESE ARE YOUR INPUT DATA SELECTIONS:

    1. UPDATE THE SYNCHRONOUS MACHINE INPUT DATA
    2. UPDATE THE TRANSMISSION LINE DATA
    3. UPDATE THE TRANSFORMER INPUT DATA
    4. DONE WITH INPUT DATA

ENTER YOUR SELECTION (1,2, 3 OR 4) 2
CHOICES :

    1. INITIALIZE FILE
    2. ADD A NEW LINE OR CHANGE AN EXISTING LINE
    3. REMOVE A LINE
    4. DISPLAY ONE LINE
    5. DISPLAY ALL LINES AFTER A GIVEN LINE
    6. DONE WITH TRANSMISSION LINE DATA

ENTER YOUR CHOICE (1,2,3,4,5 OR 6) : 5
ENTER THE NUMBER OF THE FIRST LINE TO BE DISPLAYED 1

USE THE Ctrl PRINT SCREEN OPTION TO PRINT THE RESULTS.

PRESS RETURN TO CONTINUE.

TRANSMISSION LINE INPUT DATA FOR SAMPLE RUN

LINE#	BUS -TO- BUS		X0 per unit	X1 per unit
1	2	4	1.2000	0.4000
2	2	4	1.2000	0.4000
3	2	5	0.6000	0.2000
4	2	5	0.6000	0.2000
5	4	5	0.3000	0.1000
6	4	5	0.3000	0.1000

REMOVE Ctrl PRINT SCREEN AND THEN PRESS RETURN TO CONTINUE
CHOICES :

    1. INITIALIZE FILE
    2. ADD A NEW LINE OR CHANGE AN EXISTING LINE
    3. REMOVE A LINE
    4. DISPLAY ONE LINE
    5. DISPLAY ALL LINES AFTER A GIVEN LINE
    6. DONE WITH TRANSMISSION LINE DATA

ENTER YOUR CHOICE (1,2,3,4,5 OR 6) : 6
THESE ARE YOUR INPUT DATA SELECTIONS:

    1. UPDATE THE SYNCHRONOUS MACHINE INPUT DATA
    2. UPDATE THE TRANSMISSION LINE DATA
    3. UPDATE THE TRANSFORMER INPUT DATA
    4. DONE WITH INPUT DATA

ENTER YOUR SELECTION (1,2, 3 OR 4) 3
CHOICES :

    1. INITIALIZE FILE
    2. ADD A NEW TRANSFORMER OR CHANGE AN EXISTING TRANSFORMER
    3. REMOVE A TRANSFORMER
    4. DISPLAY ONE TRANSFORMER
    5. DISPLAY ALL TRANSFORMERS AFTER A GIVEN ONE
    6. DONE WITH TRANSFORMER DATA

ENTER YOUR CHOICE (1,2,3,4,5 OR 6) : 5
ENTER THE NUMBER OF THE FIRST TRANSFORMER TO BE DISPLAYED 1

USE THE Ctrl PRINT SCREEN OPTION IF YOU WANT TO PRINT THE RESULTS.

PRESS RETURN TO CONTINUE.

TRANSFORMER INPUT DATA FOR SAMPLE RUN

TRANSF#	LOW (CNC) VOLT. BUS		HIGH(CNC) VOLT. BUS		XL	NEUTRAL REACTANCES LV	HV
					per unit	per unit	per unit
1	1	(D)	5	(Y)	0.0800		0.0000
2	3	(D)	4	(Y)	0.0400		0.0000

REMOVE Ctrl PRINT SCREEN AND THEN PRESS RETURN TO CONTINUE
CHOICES :

    1. INITIALIZE FILE
    2. ADD A NEW TRANSFORMER OR CHANGE AN EXISTING TRANSFORMER
    3. REMOVE A TRANSFORMER
    4. DISPLAY ONE TRANSFORMER
    5. DISPLAY ALL TRANSFORMERS AFTER A GIVEN ONE
    6. DONE WITH TRANSFORMER DATA

ENTER YOUR CHOICE (1,2,3,4,5 OR 6) : 6
THESE ARE YOUR INPUT DATA SELECTIONS:

    1. UPDATE THE SYNCHRONOUS MACHINE INPUT DATA
    2. UPDATE THE TRANSMISSION LINE DATA
    3. UPDATE THE TRANSFORMER INPUT DATA
    4. DONE WITH INPUT DATA

ENTER YOUR SELECTION (1,2, 3 OR 4) 4
SELECTIONS:

    1. UPDATE THE INPUT DATA
    2. RUN THE PROGRAM
    3. STOP

ENTER YOUR SELECTION (1,2 OR 3) 2

ONE OF THE FOLLOWING FAULT TYPES IS TO BE SELECTED:
    (1) THREE-PHASE FAULT
    (2) SINGLE LINE-TO-GROUND FAULT THROUGH ZF
    (3) LINE-TO-LINE FAULT THROUGH ZF
    (4) DOUBLE LINE-TO-GROUND FAULT THROUGH ZF

ENTER THE FAULT TYPE (1,2,3 OR 4)4

WHAT IS THE PREFAULT VOLTAGE (per unit)? 1.0

WHAT IS THE RESISTIVE PART OF THE FAULT IMPEDANCE(PER UNIT)? 0

WHAT IS THE REACTIVE PART OF THE FAULT IMPEDANCE(PER UNIT)? 0

DISPLAY THE FAULT CURRENTS (Y OR N)? Y

DISPLAY THE BUS VOLTAGES DURING THE FAULT (Y OR N)? Y

DISPLAY THE BUS IMPEDANCE MATRICES (Y OR N)? Y
DO YOU WANT THE OUTPUTS PRINTED IN EXPONENTIAL FORMAT (Y OR N) ? N

DO YOU WANT TO CONTINUE (Y) OR RESET THE DATA (N) ? Y

USE THE Ctrl PRINT SCREEN OPTION NOW IF YOU WANT TO PRINT THE RESULTS.
PRESS RETURN TO CONTINUE

FAULT CURRENTS FOR SAMPLE RUN

FAULT BUS	DOUBLE LINE-TO-GROUND FAULT CURRENT (PHASE B) per unit/degrees	GEN LINE OR TRSF	BUS-TO-BUS			CONTRIBUTIONS TO FAULT CURRENT PHASE A	PHASE B	PHASE C per unit/degrees
1	11.014/138.40							
		G 1	GRND	–	1	2.0270/ 90.00	7.9257/ 127.38	7.9257/ 52.62
		T 1	5	–	1	2.0270/ –90.00	3.5718/ 163.52	3.5718/ 16.48
2	5.996/156.39							
		L 1	4	–	2	0.0159/ –90.00	1.1217/ 157.14	1.1217/ 22.86
		L 2	4	–	2	0.0159/ –90.00	1.1217/ 157.14	1.1217/ 22.86
		L 3	5	–	2	0.0159/ 90.00	1.8765/ 155.95	1.8765/ 24.05
		L 4	5	–	2	0.0159/ 90.00	1.8765/ 155.95	1.8765/ 24.05
3	14.976/144.21							
		G 2	GRND	–	3	1.2139/ 90.00	12.6103/ 139.74	12.6103/ 40.26
		T 2	4	–	3	1.2139/ –90.00	2.5978/ 166.49	2.5978/ 13.51
4	13.442/135.10							
		L 1	2	–	4	0.0630/ –90.00	0.2479/ 145.47	0.2479/ 34.53
		L 2	2	–	4	0.0630/ –90.00	0.2479/ 145.47	0.2479/ 34.53
		L 5	5	–	4	0.3779/ –90.00	1.4877/ 145.47	1.4877/ 34.53
		L 6	5	–	4	0.3779/ –90.00	1.4877/ 145.47	1.4877/ 34.53
		T 2	3	–	4	0.8819/ 90.00	10.0470/ 131.53	10.0470/ 48.47
5	10.931/139.74							
		L 3	2	–	5	0.0938/ –90.00	0.4146/ 149.68	0.4146/ 30.32
		L 4	2	–	5	0.0938/ –90.00	0.4146/ 149.68	0.4146/ 30.32
		L 5	4	–	5	0.5628/ –90.00	2.4874/ 149.68	2.4874/ 30.32
		L 6	4	–	5	0.5628/ –90.00	2.4874/ 149.68	2.4874/ 30.32
		T 1	1	–	5	1.3131/ 90.00	5.3091/ 128.86	5.3091/ 51.14

BUS VOLTAGES DURING FAULTS FOR SAMPLE RUN

		PHASE A	PHASE B	PHASE C
FAULT BUS	BUS #		per unit/degrees	
1	1	0.7312/ 0.00	0.0000/ 16.10	0.0000/-16.10
	2	0.6165/ 17.60	0.3729/270.00	0.6165/162.40
	3	0.8176/ 0.00	0.6915/233.76	0.6915/126.24
	4	0.6825/ 20.85	0.4859/-90.00	0.6825/159.15
	5	0.5844/ 15.71	0.3164/270.00	0.5844/164.29
2	1	0.8274/-20.58	0.8274/200.58	0.5817/ 90.00
	2	1.1216/ 0.00	0.0000/ 8.21	0.0000/ -8.21
	3	0.8492/-22.07	0.8492/202.07	0.6382/ 90.00
	4	0.8998/ 0.00	0.5770/225.77	0.5770/134.23
	5	0.9123/ 0.00	0.4963/223.67	0.4963/136.33
3	1	0.7815/ 0.00	0.5674/226.47	0.5674/133.53
	2	0.5863/ 9.82	0.2000/270.00	0.5863/170.18
	3	0.8757/ 0.00	0.0000/180.00	0.0000/180.00
	4	0.5507/ 6.08	0.1167/270.00	0.5507/173.92
	5	0.6049/ 11.52	0.2417/270.00	0.6049/168.48
4	1	0.6049/-19.59	0.6049/199.59	0.4057/ 90.00
	2	0.6040/ 0.00	0.1406/215.53	0.1406/144.47
	3	0.5433/-16.45	0.5433/196.45	0.3077/ 90.00
	4	0.6369/ 0.00	0.0000/180.00	0.0000/180.00
	5	0.5875/ 0.00	0.2109/215.53	0.2109/144.47
5	1	0.5931/-15.03	0.5931/195.03	0.3077/ 90.00
	2	0.7420/ 0.00	0.1112/220.04	0.1112/139.96
	3	0.6886/-20.37	0.6886/200.37	0.4793/ 90.00
	4	0.6929/ 0.00	0.3337/220.04	0.3337/139.96
	5	0.7665/ 0.00	0.0000/ 0.00	0.0000/ 0.00

ZF = 0 + j 0   per unit

Zbus 0

ROW 1	0.05000	0.00000	0.00000	0.00000	0.00000
ROW 2	0.00000	0.23356	0.00000	0.01977	0.04046
ROW 3	0.00000	0.00000	0.05000	0.00000	0.00000
ROW 4	0.00000	0.01977	0.00000	0.03356	0.01287
ROW 5	0.00000	0.04046	0.00000	0.01287	0.05425

Zbus 1

ROW 1	0.10515	0.06594	0.03743	0.05406	0.07188
ROW 2	0.06594	0.15762	0.05703	0.08238	0.09525
ROW 3	0.03743	0.05703	0.07129	0.06297	0.05406
ROW 4	0.05406	0.08238	0.06297	0.09096	0.07809
ROW 5	0.07188	0.09525	0.05406	0.07809	0.10383

Zbus 2

ROW 1	0.10515	0.06594	0.03743	0.05406	0.07188
ROW 2	0.06594	0.15762	0.05703	0.08238	0.09525
ROW 3	0.03743	0.05703	0.07129	0.06297	0.05406
ROW 4	0.05406	0.08238	0.06297	0.09096	0.07809
ROW 5	0.07188	0.09525	0.05406	0.07809	0.10383

REMOVE THE Ctrl PRINT SCREEN OPTION AND PRESS RETURN TO CONTINUE
SELECTIONS:

    1. UPDATE THE INPUT DATA
    2. RUN THE PROGRAM
    3. STOP

ENTER YOUR SELECTION (1,2 OR 3) 3
Ok

9.   TRANSMISSION LINE TRANSIENTS (CHAP11)

This program computes voltage transients for the single-phase circuit shown in Figure 9a. This circuit consists of 3 single-phase lossless line sections, 10 buses, 4 independent current sources, and 10 lumped parallel RLC elements. The program computes the 10 bus voltages at discrete times Δt, $2\Delta t$, ... , $N\Delta t$.

Input data include line data, RLC data, and independent current source data, which are stored in three separate data files. The user can store up to five separate cases. The line, RLC, and source data files for Case 1 are called TLTL1.DAT, TLTRLC1.DAT, and TLTS1.DAT. Input data files for Case 2 (3,4,5) have similar names,   except the number 2 (3,4,5) is used.

The user has the following three options for setting circuit breakers B1 and B2 in Figure 9a :

(1) B1 and B2 open - The first 4 buses are considered along with line section 1, the first 4 RLC elements, and the first 2 sources. The remainder of the circuit is disregarded.

(2) B1 closed and B2 open - The first 7 buses are considered along with line sections 1 and 2, the first 7 RLC elements, and the first 3 sources. The remainder of the circuit is disregarded.

(3) B1 and B2 closed - The entire 10-bus circuit is considered.

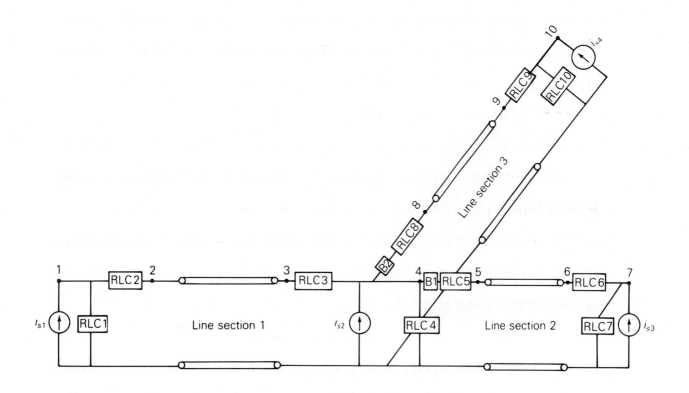

Figure 9a Circuit diagram for CHAP11: TRANSMISSION LINE
TRANSIENTS

Line input data include the characteristic impedance Z_c (Ω) and transit time tau (μs) for each lossless line section.

A lossy line section can be modeled by lumping the series line resistance and including half of this lumped resistance in the R element on each side of the line section.

RLC input data include the values of 1/R (1/ohms), 1/L (1/henries) and C (farads) for each parallel RLC element. Two values along with a "changeover voltage" (kV) are entered for each element. The program uses the first value when the voltage across the element is less than the changeover voltage, and the second value when the voltage is greater than or equal to the changeover voltage. In this way, voltage-dependent, piecewise constant RLC elements can be modelled.

The user can select an open circuit for any RLC element by entering zero values for 1/R, 1/L, and C. Also, the user can approximate a short circuit for any RLC element by entering a large value for 1/R (for example 1/R = 1000) and zero values for 1/L and C. However, do not select too high a value for 1/R , otherwise the computation method may break down.

The user has the following four options for each independent current source, as shown in Figure 9b : (S) square wave; (R) ramp; (T) triangular wave; or (E) double exponential wave. Source input data include the source type (S, R, T, or E), source magnitude I (kA or kA/s); and the times T1 and T2 (μs) for each independent current source.

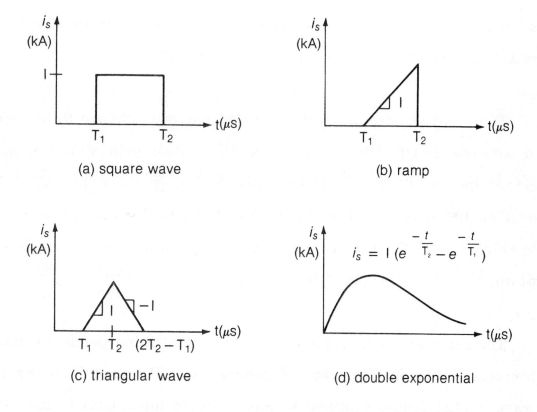

Figure 9b Independent current sources for CHAP11

The user has the following three options: (1) update the input data files; (2) run the program; or (3) stop.

When the user updates the input data files, the following options for each input data file are given: (1) create a new file; (2) change the data for one line section, RLC element, or source; or (3) display the file.

When the user runs the program, the calculation time interval Δt (μs) and the final time t_F (μs) are selected. Δt should be small in order to obtain an accurate solution. One guideline is to select Δt to be 1/10 of either the smallest transit time tau or the smallest time constant of your circuit. However, a program restriction is that the ratio tau/Δt must be less than 250 for each line section. That is, the program will not allow too small a value of Δt.

The program replaces the circuit elements in Figure 9a by their discrete-time equivalent circuits. Then nodal analysis is used to compute the 10 bus voltages at discrete time intervals Δt, $2\Delta t$, ... , until $N\Delta t = t_F$. Gauss elimination is used to solve the nodal equations.

The user selects up to 5 bus voltages for display during any run. The user also selects the printout integer K. Then the bus voltages at time intervals of $K\Delta t$ are displayed. Also, the user can display the bus voltages in exponential format (scientific notation) or fixed-point format (numbers with decimal points) .

The following sample run solves Example 11.10 of the text *Power System Analysis and Design with Personal Computer Applications,* by J.D. Glover and M. Sarma.

CHAP11   SAMPLE RUN   EXAMPLE 11.10

```
ENTER YOUR CASE NAME EXAMPLE 11.10

IS CIRCUIT BREAKER B1 CLOSED (Y OR N) ? y
IS CIRCUIT BREAKER B2 CLOSED (Y OR N) ? n

WOULD YOU LIKE TO:
        1. CREATE, EDIT OR LOOK AT THE INPUT DATA FILES
        2. RUN THE PROGRAM
        3. STOP

ENTER YOUR SELECTION(1,2, OR 3) 2
ENTER THE CALCULATION TIME INTERVAL (MICROSECONDS) 0.1

ENTER THE FINAL TIME (MICROSECONDS) 150

YOU MAY PRINT FROM  1 TO  5 OUTPUTS
ENTER THE NUMBER OF OUTPUTS 3
ENTER OUTPUT   1   BUS NUMBER
? 1
ENTER OUTPUT   2   BUS NUMBER
? 4
ENTER OUTPUT   3   BUS NUMBER
? 7
THE OUTPUTS ARE PRINTED EVERY KTH TIME INTERVAL

ENTER THE PRINTOUT INTEGER K(ANY INTEGER>=1)30

DO YOU WANT THE OUTPUTS PRINTED IN EXPONENTIAL FORMAT (Y OR N )? n

DO YOU WANT TO PRINT THE INPUT DATA (Y OR N) ? y

DO YOU WANT TO CONTINUE (Y) OR RESET THE DATA (N) ? y

USE THE Ctrl PRINT SCREEN NOW IF YOU WANT A PRINTED OUTPUT

PRESS RETURN TO CONTINUE
```

INPUT DATA FOR EXAMPLE 11.10

LINE DATA FOR EXAMPLE 11.10

LINE SECTION	CHARACTERISTIC IMPEDANCE	TRANSIT TIME
	ohms	micro seconds
1	300	16.67
2	300	16.67

CIRCUIT BREAKER ONE IS CLOSED

CIRCUIT BREAKER TWO IS OPEN

RLC DATA FOR EXAMPLE 11.10

RLC	1/R			1/L			C		
ELEMENT	FIRST VALUE	SECOND VALUE	CHANGE VOLTAGE	FIRST VALUE	SECOND VALUE	CHANGE VOLTAGE	FIRST VALUE	SECOND VALUE	CHANGE VOLTAGE
	1/ohms	1/ohms	kV	1/henry	1/henry	kV	farads	farads	kV
1	0.0E+00	0.0E+00	0.0E+00	0.0E+00	0.0E+00	0.0E+00	0.0E+00	0.0E+00	0.0E+00
2	8.0E+00	8.0E+00	0.0E+00	0.0E+00	0.0E+00	0.0E+00	0.0E+00	0.0E+00	0.0E+00
3	8.0E+00	8.0E+00	0.0E+00	0.0E+00	0.0E+00	0.0E+00	0.0E+00	0.0E+00	0.0E+00
4	0.0E+00	0.0E+00	0.0E+00	0.0E+00	0.0E+00	0.0E+00	0.0E+00	0.0E+00	0.0E+00
5	8.0E+00	8.0E+00	0.0E+00	0.0E+00	0.0E+00	0.0E+00	0.0E+00	0.0E+00	0.0E+00
6	8.0E+00	8.0E+00	0.0E+00	0.0E+00	0.0E+00	0.0E+00	0.0E+00	0.0E+00	0.0E+00
7	5.0E-07	2.2E-01	5.5E+01	0.0E+00	0.0E+00	0.0E+00	6.0E-09	6.0E-09	0.0E+00

INDEPENDENT CURRENT SOURCE DATA FOR EXAMPLE 11.10

SOURCE	TYPE	I	T1	T2
		kA or kA/S	micro seconds	micro seconds
1	S	0	0	0
2	S	20	0	20
3	S	0	0	0

OUTPUT DATA FOR EXAMPLE 11.10

BUS VOLTAGES

TIME micro seconds	V 1 kV	V 4 kV	V 7 kV
0.000	0.00	3001.29	0.00
3.000	0.00	3001.29	0.00
6.000	0.00	3001.29	0.00
9.000	0.00	3001.29	0.00
12.000	0.00	3001.29	0.00
15.000	0.00	3001.29	0.00
18.000	6000.52	3001.29	88.64
21.000	6000.52	0.00	88.64
24.000	6000.52	0.00	88.64
27.000	6000.52	0.00	88.64
30.000	6000.52	0.00	88.64
33.000	6000.52	0.00	88.64
36.000	6000.52	91.54	88.64
39.000	0.00	91.54	-7.34
42.000	0.00	91.54	-1.39
45.000	0.00	91.54	-0.26
48.000	0.00	91.54	-0.05
51.000	-5813.37	91.54	88.58
54.000	-5813.37	-17.85	88.58
57.000	-5813.37	-3.37	88.58
60.000	-5813.37	-0.64	88.58
63.000	-5813.37	-0.12	88.58
66.000	-5813.37	-0.02	88.58
69.000	-5813.37	-5813.95	88.58
72.000	-16.39	-5813.95	-8.21
75.000	-3.10	-5813.95	-1.55
78.000	-0.58	-5813.95	-0.29
81.000	-0.11	-5813.95	-0.06
84.001	-5814.53	-5813.95	-85.89
87.001	-5814.53	-39.87	-85.89
90.001	-5814.53	-7.54	-85.89
93.001	-5814.53	-1.42	-85.89
96.001	-5814.53	-0.27	-85.89
99.001	-5814.53	-0.05	-85.89
102.001	-5814.53	-88.78	-85.89
105.001	-18.32	-88.78	-10.56
108.001	-3.47	-88.78	-7.76
111.001	-0.66	-88.78	-2.56
114.001	-0.12	-88.78	-0.69
117.001	5633.12	-88.78	-85.81
120.001	5633.03	13.94	-85.84
123.001	5633.03	-11.40	-85.84
126.001	5633.03	-4.80	-85.84
129.001	5633.03	-1.41	-85.84
132.001	5633.03	-0.36	-85.84
135.001	5633.03	5633.67	-85.84
138.001	1.41	5633.67	-9.51
141.001	-12.62	5633.67	-8.25
144.001	-4.82	5633.67	-2.78
147.001	-1.37	5633.67	-0.76

10 TRANSIENT STABILITY (CHAP12)

This program computes machine power angles and frequencies in a three-phase power system subjected to disturbances. The program also computes machine angular accelerations, machine electrical power outputs, and bus voltage magnitudes.

Before running this program, it is first necessary to run CHAP7, POWER FLOW, for your power system under consideration. CHAP12 opens and then reads data from the CHAP7 input/output data files.

Input data for CHAP12 also include synchronous machine data, which are stored in a data file. The user can store up to three separate cases.   The synchronous machine data file for Case 1 is called TSM1.DAT. The data file for Case 2 (3) has a similar name, except the number 2 (3) is used. The maximum number of machines is 25.

Synchronous machine data include the machine bus number, the per-unit transient reactance, and the per-unit H constant for each machine.

The user has the following options: (1) update the synchronous machine input data; (2) set the disturbances and run the program; or (3) stop.

When the user updates the synchronous machine input data, the following options are given: (1) initialize the file (which removes all data currently stored); (2) add a new machine or change an existing one; (3) remove a machine; (4) display the data for one machine; and (5) display the data for all machines after a given one.

When the user sets the disturbances and runs the program, the integration interval Δt (s) and final time t_F (s) are first selected. Δt should be small in order that an accurate solution is obtained. Typical values are $\Delta t = 0.01$ s or $\Delta t = 0.01667$ s (one cycle in a 60 Hz system).

Next the user selects the number of disturbances (1,2,3, or 4) and the disturbance times (s). The user has the following options for the first disturbance: (1) put a three-phase short circuit at a bus; and (2) open one or more machine breakers, transmission line breakers, or transformer breakers.

The user has the following disturbance options after the first disturbance: (1) put a three-phase short circuit at a bus; (2) extinguish an existing short circuit; (3) open one or more machine breakers, transmission line breakers, or transformer breakers; and (4) close one or more machine breakers, transmission line breakers, or transformer breakers .

The program alternately solves, step-by-step, algebraic power flow equations representing the network and differential equations representing the machines. The Gauss-Seidel iterative method is used to solve the power flow equations, and a modified Euler's method is used to solve the differential equations.

The user can select from the following outputs: (1) machine power angle; (2) machine frequency; (3) machine real power output; and (4) bus voltage. Up to five outputs can be displayed during any run. The user also selects the printout integer K. Then the outputs are displayed at intervals of $K\Delta t$. The outputs can be displayed in exponential format (scientific notation) or in fixed-point format (numbers with decimal points).

The following sample run is for a temporary three-phase short circuit on line 6 at bus 5 of the power system whose single-line diagram is shown in Figure10a. Note that this is the same single-line diagram as that in Figure 6a. For prefault conditions, the power flow output data from the CHAP7 sample run are used. Machine input data are given in the CHAP12 sample run. The short circuit is cleared by opening circuit breakers B42 and B52 at t = 0.05 s (3 cycles), followed by reclosing these breakers at t = 0.50 s (30 cycles after fault clearing). When reclosure occurs, the temporary fault has already been extinguished. The integration interval is Δt = 0.01 s and the final time is t_F = 0.75 s.

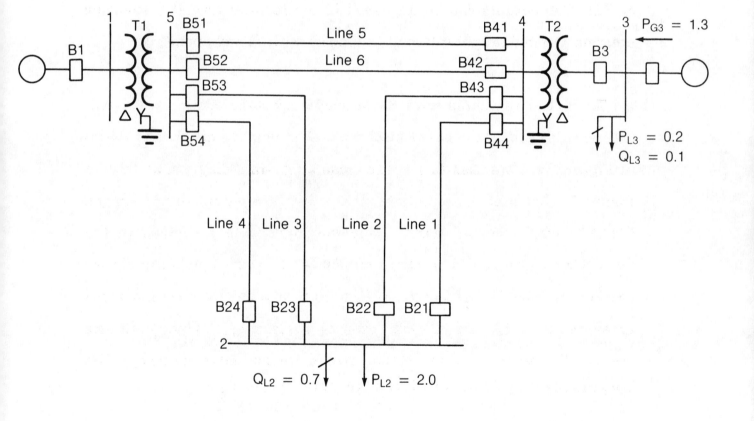

Figure 10a Single-line diagram for SAMPLE RUN : CHAP12

(per-unit values are shown)

WHAT IS THE CASE NUMBER FOR YOUR POWER FLOW PROGRAM?
(ENTER AN INTEGER FROM 1 TO 3) 1

ENTER YOUR CASE NAME SAMPLE RUN

WOULD YOU LIKE TO:
       1. UPDATE THE SYNCHRONOUS MACHINE INPUT DATA
       2. SET DISTURBANCES AND RUN THE PROGRAM
       3. STOP

ENTER YOUR SELECTION (1,2 OR 3) 1

SELECTIONS:
  1. INITIALIZE FILE
  2. ADD A NEW SYNCHRONOUS MACHINE OR CHANGE AN EXISTING ONE
  3. REMOVE A MACHINE
  4. DISPLAY ONE MACHINE
  5. DISPLAY ALL MACHINES BELOW A GIVEN ONE
  6. DONE WITH MACHINE DATA

ENTER YOUR SELECTION (1 TO 6) 5
ENTER THE NUMBER OF THE FIRST MACHINE TO BE DISPLAYED 1

USE THE Ctrl PRINT SCREEN OPTION NOW IF YOU WANT TO PRINT THE RESULTS.

PRESS RETURN TO CONTINUE.
      SYNCHRONOUS MACHINE DATA FOR SAMPLE RUN

MACHINE#	BUS	X' per unit	H per unit seconds
1	1	.2	5
2	3	.1	50

REMOVE Ctrl PRINT SCREEN AND THEN PRESS RETURN TO CONTINUE

SELECTIONS:
  1. INITIALIZE FILE
  2. ADD A NEW SYNCHRONOUS MACHINE OR CHANGE AN EXISTING ONE
  3. REMOVE A MACHINE
  4. DISPLAY ONE MACHINE
  5. DISPLAY ALL MACHINES BELOW A GIVEN ONE
  6. DONE WITH MACHINE DATA

ENTER YOUR SELECTION (1 TO 6) 6

WOULD YOU LIKE TO:
       1. UPDATE THE SYNCHRONOUS MACHINE INPUT DATA
       2. SET DISTURBANCES AND RUN THE PROGRAM
       3. STOP

```
ENTER YOUR SELECTION ( 1,2 OR 3 ) 2


ENTER THE INTEGRATION TIME INTERVAL (SECONDS)  0.01
ENTER THE FINAL TIME (SECONDS)  0.760
ENTER THE NUMBER OF DISTURBANCE TIMES (1,2,3 OR 4) 3

FIRST DISTURBANCE
ENTER THE FIRST DISTURBANCE TIME(SECONDS) 0
IS THERE A SHORT CIRCUIT AT THE FIRST DISTURBANCE TIME(Y OR N) ? Y
ENTER THE SHORT CIRCUIT BUS NUMBER 5
DO GENERATOR BREAKERS OPEN AT THE FIRST DISTURBANCE(Y OR N) ? N
DO LINE   BREAKERS OPEN AT THE FIRST DISTURBANCE(Y OR N) ? N
DO TRANSFORMER  BREAKERS OPEN AT THE FIRST DISTURBANCE(Y OR N) ? N

SECOND DISTURBANCE
ENTER THE SECOND DISTURBANCE TIME(SECONDS) 0.05
IS THERE A SHORT CIRCUIT AT THE SECOND DISTURBANCE TIME(Y OR N) ? N
IS THE SHORT CIRCUIT AT THE FIRST DISTURBANCE EXTINGUISHED (Y OR N) ? Y
DO  GENERATOR  BREAKERS OPEN AT THE SECOND DISTURBANCE(Y OR N) ? N
DO LINE   BREAKERS OPEN AT THE SECOND DISTURBANCE(Y OR N) ? Y
HOW MANY LINE   BREAKERS OPEN ? 2
TRANSMISSION LINE OPENING CIRCUIT BREAKER 1
ENTER THE LINE NUMBER 6
ENTER THE BUS NUMBER 4
TRANSMISSION LINE OPENING CIRCUIT BREAKER 2
ENTER THE LINE NUMBER 6
ENTER THE BUS NUMBER 5
DO  TRANSFORMER  BREAKERS OPEN AT THE SECOND DISTURBANCE(Y OR N) ? N

THIRD DISTURBANCE
ENTER THE THIRD DISTURBANCE TIME(SECONDS) 0.50
IS THERE A SHORT CIRCUIT AT THE THIRD DISTURBANCE TIME(Y OR N) ? N
DO  GENERATOR  BREAKERS OPEN AT THE THIRD DISTURBANCE(Y OR N) ? N
DO  GENERATOR  BREAKERS RECLOSE AT THE THIRD DISTURBANCE(Y OR N) ? N
DO LINE   BREAKERS OPEN AT THE THIRD DISTURBANCE(Y OR N) ? N
DO  LINE   BREAKERS RECLOSE AT THE THIRD DISTURBANCE(Y OR N) ? Y
HOW MANY   LINE   BREAKERS CLOSE  ? 2
TRANSMISSION LINE CLOSING CIRCUIT BREAKER 1
ENTER THE LINE NUMBER 6
ENTER THE BUS NUMBER 4
TRANSMISSION LINE CLOSING CIRCUIT BREAKER 2
ENTER THE LINE NUMBER 6
ENTER THE BUS NUMBER 5
DO  TRANSFORMER  BREAKERS OPEN AT THE THIRD  DISTURBANCE(Y OR N) ? N
DO  TRANSFORMER  BREAKERS RECLOSE AT THE THIRD DISTURBANCE(Y OR N) ? N

FOR OUTPUT LABELLING
ENTER THE FIRST DISTURBANCE NAME(ANY ALPHANUMERIC SEQUENCE) FAULT AT BUS 5

ENTER THE SECOND DISTURBANCE NAME                FAULT CLEARED
ENTER THE THIRD DISTURBANCE NAME RECLOSURE
```

```
YOU MAY PRINT FROM 1 TO 5 OUTPUTS
ENTER THE NUMBER OF OUTPUTS ( 1 TO 5 ) 5

OUTPUT   1
     SELECTIONS:
                1.  MACHINE POWER ANGLE
                2.  MACHINE FREQUENCY
                3.  MACHINE REAL POWER OUTPUT
                4.  BUS VOLTAGE

ENTER YOUR SELECTION (1,2,3 OR 4) 1
ENTER THE MACHINE NUMBER 1

OUTPUT   2
     SELECTIONS:
                1.  MACHINE POWER ANGLE
                2.  MACHINE FREQUENCY
                3.  MACHINE REAL POWER OUTPUT
                4.  BUS VOLTAGE

ENTER YOUR SELECTION (1,2,3 OR 4) 2
ENTER THE MACHINE NUMBER 1

OUTPUT   3
     SELECTIONS:
                1.  MACHINE POWER ANGLE
                2.  MACHINE FREQUENCY
                3.  MACHINE REAL POWER OUTPUT
                4.  BUS VOLTAGE

ENTER YOUR SELECTION (1,2,3 OR 4) 1
ENTER THE MACHINE NUMBER 2

OUTPUT   4
     SELECTIONS:
                1.  MACHINE POWER ANGLE
                2.  MACHINE FREQUENCY
                3.  MACHINE REAL POWER OUTPUT
                4.  BUS VOLTAGE

ENTER YOUR SELECTION (1,2,3 OR 4) 2
ENTER THE MACHINE NUMBER 2

OUTPUT   5
     SELECTIONS:
                1.  MACHINE POWER ANGLE
                2.  MACHINE FREQUENCY
                3.  MACHINE REAL POWER OUTPUT
                4.  BUS VOLTAGE

ENTER YOUR SELECTION (1,2,3 OR 4) 4
ENTER THE BUS NUMBER 5
THE OUTPUTS ARE PRINTED EVERY KTH  TIME INTERVAL
ENTER THE PRINTOUT INTEGER K  ( K >= 1 ) 3
DO YOU WANT THE OUTPUTS IN EXPONENTIAL FORMAT (Y OR N) ? N
CONTINUE ( Y ) OR RESET THE DISTURBANCES (N) ? Y
USE THE Ctrl PRINT SCREEN OPTION NOW IF YOU WANT TO PRINT THE OUTPUTS
```

TRANSIENT STABILITY OUTPUT FOR SAMPLE RUN

TIME	DELTA 1	OMEGA 1	DELTA 2	OMEGA 2	V 5
seconds	degrees	rad/s	degrees	rad/s	per unit

FAULT AT BUS 5

0.000	12.72	376.99	5.77	376.99	0.000
0.030	13.58	377.99	5.86	377.11	0.000

FAULT CLEARED

0.060	16.14	378.98	6.16	377.22	0.999
0.090	19.45	378.83	6.58	377.24	0.996
0.120	22.41	378.58	7.04	377.28	0.994
0.150	24.87	378.25	7.56	377.32	0.991
0.180	26.73	377.87	8.16	377.36	0.989
0.210	27.90	377.47	8.84	377.41	0.988
0.240	28.36	377.06	9.61	377.46	0.989
0.270	28.13	376.67	10.47	377.51	0.990
0.300	27.28	376.33	11.40	377.55	0.992
0.330	25.89	376.06	12.40	377.59	0.995
0.360	24.13	375.88	13.45	377.61	0.998
0.390	22.14	375.82	14.53	377.63	1.001
0.420	20.14	375.86	15.63	377.63	1.003
0.450	18.30	376.01	16.72	377.62	1.004
0.480	16.82	376.26	17.79	377.60	1.004

RECLOSURE

0.510	15.84	376.60	18.82	377.57	1.022
0.540	15.52	377.02	19.78	377.53	1.024
0.570	15.96	377.48	20.66	377.48	1.024
0.600	17.19	377.93	21.45	377.43	1.024
0.630	19.18	378.35	22.17	377.38	1.024
0.660	21.83	378.70	22.81	377.35	1.024
0.690	25.01	378.96	23.39	377.32	1.024
0.720	28.53	379.10	23.93	377.30	1.023
0.750	32.21	379.13	24.45	377.29	1.021

REMOVE Ctrl PRINT SCREEN AND THEN PRESS RETURN TO CONTINUE

WOULD YOU LIKE TO:
      1. UPDATE THE SYNCHRONOUS MACHINE INPUT DATA
      2. SET DISTURBANCES AND RUN THE PROGRAM
      3. STOP

ENTER YOUR SELECTION (1,2 OR 3) 3

Figure 10b shows the single-line diagram of a six-bus power system, which is to be used for working Problems 12.16, 12.17, and 12.23 in the text *Power System Analysis and Design With Personal Computer Applications* , by J.D. Glover and M. Sarma.

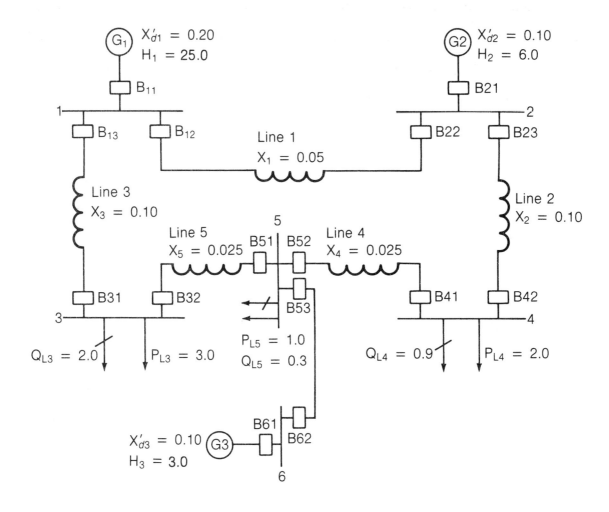

Figure 10b Single-line diagram of a six-bus power system

(per-unit values are shown)

11. PROGRAM LISTINGS AND FLOWCHARTS

This section is primarily intended for those readers who are interested in writing their own computer programs. Program listings of the nine subroutines in the program CHAP2 and flow charts for the remaining programs are given.

Background and theory for the programs are given in the text *Power System Analysis and Design with Personal Computer Applications* , by J.D. Glover and M. Sarma, PWS Publishers, Boston, 1987. Equation numbers from the text are listed in the the flowcharts given here.

```
11000 '*********SUBROUTINE RMA(A,B,C,N,M)***********
11010 'THIS SUBROUTINE COMPUTES THE MATRIX SUM C = A + B
11020 'OF THE TWO NxM MATRICES A AND B
11030 '*****************************************************
11040 FOR I=1 TO N
11050 FOR J=1 TO M
11060 C(I,J)=A(I,J)+B(I,J)
11070 NEXT J
11080 NEXT I
11090 RETURN
11100 '************END OF RMA SUBROUTINE*****************
12000 '********SUBROUTINE CMA(A,B,C,N,M)***************
12010 'THIS SUBROUTINE COMPUTES THE MATRIX SUM C = A + B
12020 ' OF THE TWO NxM COMPLEX MATRICES A AND B
12030 '*****************************************************
12040 FOR I=1 TO N
12050 FOR J=1 TO M
12060 CR(I,J)=AR(I,J)+BR(I,J)
12070 CI(I,J)=AI(I,J)+BI(I,J)
12080 NEXT J
12090 NEXT I
12100 RETURN
12110 '************END OF CMA SUBROUTINE*****************
13000 '**********SUBROUTINE RMM(A,B,C,N,M,P)***********
13010 'THIS SUBROUTINE COMPUTES THE MATRIC PRODUCT C = AB
13020 'OF THE NxM REAL MATRIX A AND THE MxP REAL MATRIX B.
13030 'THE RESULTING MATRIX C HAS DIMENSION NxP
13040 '********************************************************
13050 FOR I=1 TO N
13060 FOR J=1 TO P
13070 C(I,J)=0
13080 NEXT J
13090 NEXT I
13100 FOR I=1 TO N
13110 FOR J=1 TO P
13120 FOR K=1 TO M
13130 C(I,J)=C(I,J)+A(I,K)*B(K,J)
13140 NEXT K
13150 NEXT J
13160 NEXT I
13170 RETURN
13180 '**********END OF RMM SUBROUTINE*********************
```

```
14000 '**************SUBROUTINE CMM(A,B,C,N,M,P)**********
14010 'THIS SUBROUTINE COMPUTES THE MATRIX PRODUCT C = A B
14020 'OF THE TWO NxM COMPLEX MATRICES A AND B
14030 'THE RESULTING COMPLEX MATRIX C HAS DIMENSION NxP
14040 '*********************************************************
14050 FOR I=1 TO N
14060 FOR J=1 TO P
14070 CR(I,J)=0
14080 CI(I,J)=0
14090 NEXT J
14100 NEXT I
14110 FOR I=1 TO N
14120 FOR J=1 TO P
14130 FOR K=1 TO M
14140 CR(I,J)=CR(I,J)+AR(I,K)*BR(K,J)-AI(I,K)*BI(K,J)
14150 CI(I,J)=CI(I,J)+AR(I,K)*BI(K,J)+AI(I,K)*BR(K,J)
14160 NEXT K
14170 NEXT J
14180 NEXT I
14190 RETURN
14200 '************END OF CMM SUBROUTINE********************


15000 '***********SUBROUTINE RMI(A,N)*********************
15010 'THIS SUBROUTINE COMPUTES THE INVERSE OF
15020 'THE NxN REAL MATRIX A,WHOSE DETERMINANT IS
15030 'ASSUMED TO BE NONZERO.
15040 'THE GAUSS ELIMINATION METHOD IS EMPLOYED.
15050 'IT IS ALSO ASSUMED THAT THERE IS NO ZERO PIVOT ELEMENT.
15060 'THE INVERSE MATRIX IS CALLED A AND IS STORED IN THE SAME
15070 'LOCATION AS THE ORIGINAL MATRIX. THE ORIGINAL A MATRIX
15080 'IS ERASED.
15090 FOR I=1 TO N
15100 IF A(I,I)<>0 THEN GOTO 15130
15110 PRINT " THE GAUSS ELIMINATION METHOD GIVES A ZERO PIVOT ELEMENT. "
15112 PRINT " PLEASE CHANGE YOUR MATRIX."
15120 GOTO 5230
15130 A(I,I)=1/A(I,I)
15140 FOR J=1 TO N
15150 IF J=I GOTO 15230
15160 A(J,I)=A(J,I)*A(I,I)
15170 FOR K=1 TO N
15180 IF K=I GOTO 15220
15190 A(J,K)=A(J,K)-A(J,I)*A(I,K)
15200 IF J=N THEN 15210 ELSE 15220
15210 A(I,K)=-A(I,I)*A(I,K)
15220 NEXT K
15230 NEXT J
15240 NEXT I
15250 K=N-1
15260 FOR J=1 TO K
15270 A(N,J)=-A(N,N)*A(N,J)
15280 NEXT J
15290 RETURN
15300 '************END OF RMI SUBROUTINE********************
```

```
16000 '*********SUBROUTINE CMI(A,N)******************************
16010 'THIS SUBROUTINE COMPUTES THE INVERSE OF
16020 'THE NxN COMPLEX MATRIX A = AR +jAI, WHOSE DETERMINANT IS
16030 'ASSUMED TO BE NONZERO.
16040 'THE GAUSS ELIMINATION METHOD IS USED.
16050 'IT IS ALSO ASSUMED THAT THERE IS NO ZERO PIVOT ELEMENT.
16060 'THE INVERSE MATRIX IS CALLED A AND IS STORED IN THE
16070 'LOCATION AS THE ORIGINAL MATRIX. THE ORIGINAL
16080 'A MATRIX IS THEREFORE ERASED.
16090 FOR I=1 TO N
16100 ASQ=AR(I,I)^2+AI(I,I)^2
16110 IF ASQ<>0 THEN GOTO 16140
16120 PRINT " THE GAUSS ELIMINATION METHOD GIVES A ZERO PIVOT ELEMENT."
16122 PRINT " PLEASE CHANGE YOUR MATRIX."
16130 GOTO 6560
16140 AR(I,I)=AR(I,I)/ASQ
16150 AI(I,I)=-AI(I,I)/ASQ
16160 FOR J=1 TO N
16170 IF J=I GOTO 16340
16180 BBB=AR(J,I)*AR(I,I)-AI(J,I)*AI(I,I)
16190 CCC=AR(J,I)*AI(I,I)+AI(J,I)*AR(I,I)
16200 AR(J,I)=BBB
16210 AI(J,I)=CCC
16220 FOR K=1 TO N
16230 IF K=I GOTO 16330
16240 DDD=AR(J,K)-AR(J,I)*AR(I,K)+AI(J,I)*AI(I,K)
16250 EEE=AI(J,K)-AR(J,I)*AI(I,K)-AI(J,I)*AR(I,K)
16260 AR(J,K)=DDD
16270 AI(J,K)=EEE
16280 IF J=N THEN 16290 ELSE 16330
16290 FFF=-AR(I,I)*AR(I,K)+AI(I,I)*AI(I,K)
16300 GGG=-AR(I,I)*AI(I,K)-AI(I,I)*AR(I,K)
16310 AR(I,K)=FFF
16320 AI(I,K)=GGG
16330 NEXT K
16340 NEXT J
16350 NEXT I
16360 K=N-1
16370 FOR J=1 TO K
16380 HHH=-AR(N,N)*AR(N,J)+AI(N,N)*AI(N,J)
16390 PPP=-AR(N,N)*AI(N,J)-AI(N,N)*AR(N,J)
16400 AR(N,J)=HHH
16410 AI(N,J)=PPP
16420 NEXT J
16430 RETURN
16440 '*************END OF CMI SUBROUTINE********************

17000 '**********SUBROUTINE RMT(A,AT,N,M)*******************
17010 'THIS SUBROUTINE COMPUTES THE MATRIX TRANSPOSE A T
17015 ' OF THE REAL MATRIX A.
17020 FOR I=1 TO M
17030 FOR J=1 TO N
17040 AT(I,J)=A(J,I)
17050 NEXT J
17060 NEXT I
17070 RETURN
17080 '*******END OF RMT SUBROUTINE***********************
```

```
18000 '*********SUBROUTINE CMT(A,AT,N,M)*********************
18010 'THIS SUBROUTINE COMPUTES THE MATRIX TRANSPOSE AT
18020 'OF THE COMPLEX MATRIX A.
18030 FOR I=1 TO M
18040 FOR J=1 TO N
18050 ATR(I,J)=AR(J,I)
18060 ATI(I,J)=AI(J,I)
18070 NEXT J
18080 NEXT I
18090 RETURN
18100 '**************END OF CMT SUBROUTINE*********************
```

```
19000 '**********SUBROUTINE CMC(A,AC,N,M)**********************
19010 'THIS SUBROUTINE COMPUTES THE MATRIX CONJUGATE AC
19020 'OF THE COMPLEX MATRIX A.
19030 FOR I=1 TO N
19040 FOR J=1 TO M
19050 ACR(I,J)=AR(I,J)
19060 ACI(I,J)=-AI(I,J)
19070 NEXT J
19080 NEXT I
19090 RETURN
19100 '*********END OF CMC SUBROUTINE*************************
```

CHAP3 SYMMETRICAL COMPONENTS

Selections:

```
┌─────────────────┐
│  1. SEQVEC      │ ◄──────── Return
│  2. PHASVEC     │
│  3. SEQIMP      │
│  4. EXIT        │
└─────────────────┘
```

1   2   3

```
┌──────────────┐   ┌──────────────┐   ┌──────────────┐
│  Enter  V_p  │   │  Enter  V_s  │   │  Enter  Z_p  │
└──────────────┘   └──────────────┘   └──────────────┘

┌──────────────┐   ┌──────────────┐   ┌──────────────┐
│ Compute  V_s │   │ Compute  V_p │   │ Compute  Z_s │
│   (3.1.11)   │   │   (3.1.9)    │   │   (3.2.9)    │
└──────────────┘   └──────────────┘   └──────────────┘

┌──────────────┐   ┌──────────────┐   ┌──────────────┐
│Display V_p,V_s│  │Display V_s,V_p│  │Display Z_p,Z_s│
└──────────────┘   └──────────────┘   └──────────────┘
```

Return

CHAP5   LINE CONSTANTS

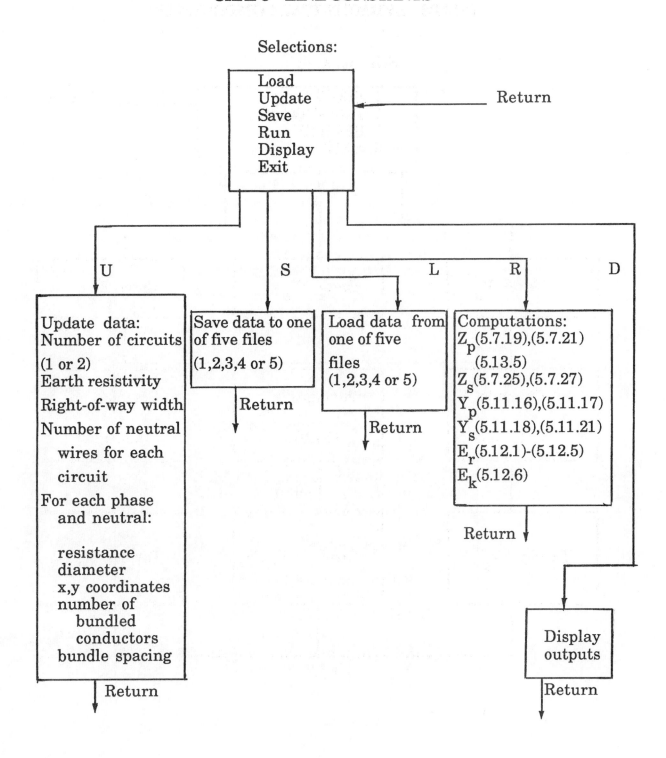

Selections:

Load
Update
Save
Run
Display
Exit

Return

U   S   L   R   D

Update data:
Number of circuits
(1 or 2)
Earth resistivity
Right-of-way width
Number of neutral
  wires for each
  circuit
For each phase
  and neutral:

  resistance
  diameter
  x,y coordinates
  number of
    bundled
    conductors
  bundle spacing

Return

Save data to one
of five files
(1,2,3,4 or 5)

Return

Load data from
one of five
files
(1,2,3,4 or 5)

Return

Computations:
Z_p(5.7.19),(5.7.21)
  (5.13.5)
Z_s(5.7.25),(5.7.27)
Y_p(5.11.16),(5.11.17)
Y_s(5.11.18),(5.11.21)
E_r(5.12.1)-(5.12.5)
E_k(5.12.6)

Return

Display
outputs

Return

Enter the input data:

```
Name of line
Three-phase or single-phase
Rated line voltage
Line length
Series resistance and reactance
Shunt conductance and susceptance
Number (0,1 or 2 ) and location of
    intermediate substations
% series and shunt compensation
    at each line end and at
    each intermediate substation
```
→ Return

Computations:

Without compensation :

Characteristic impedance (6.2.16)
Propagation constant (6.2.12)
Wavelength (6.4.15)
Surge impedance loading (6.4.21)
Equivalent π circuit (6.3.5)-(6.3.10)
ABCD parameters (6.3.1)-(6.3.3)

With compensation :

ABCD parameters of each series and
    shunt compensation network (Fig.6.4)
Equivalent ABCD parameters of
    the compensated line (Fig.6.4)
Sending-end quantities (6.1.3)
% Voltage regulation (6.1.8)
Maximum real power delivered (6.5.6)

Display outputs

Return

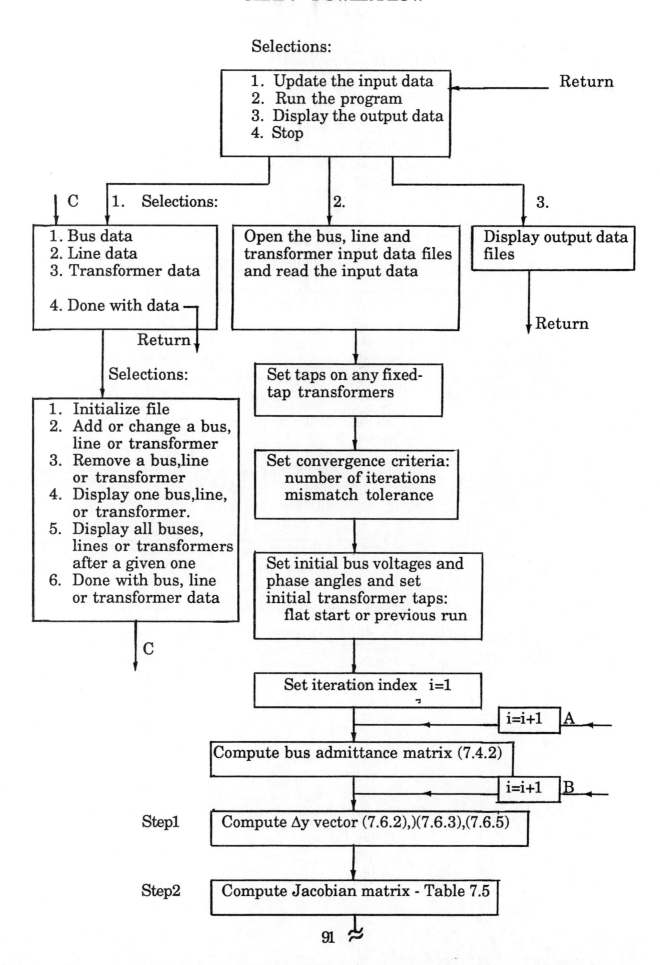

Selections:

1. Update the input data
2. Run the program
3. Display the output data
4. Stop

Return

C 1.   Selections:

1. Bus data
2. Line data
3. Transformer data

4. Done with data

Return

Selections:

1. Initialize file
2. Add or change a bus, line or transformer
3. Remove a bus,line or transformer
4. Display one bus,line, or transformer.
5. Display all buses, lines or transformers after a given one
6. Done with bus, line or transformer data

C

2.

Open the bus, line and transformer input data files and read the input data

Set taps on any fixed-tap transformers

Set convergence criteria: number of iterations mismatch tolerance

Set initial bus voltages and phase angles and set initial transformer taps: flat start or previous run

Set iteration index i=1

i=i+1 A

Compute bus admittance matrix (7.4.2)

i=i+1 B

Step1   Compute Δy vector (7.6.2),)(7.6.3),(7.6.5)

Step2   Compute Jacobian matrix - Table 7.5

3.

Display output data files

Return

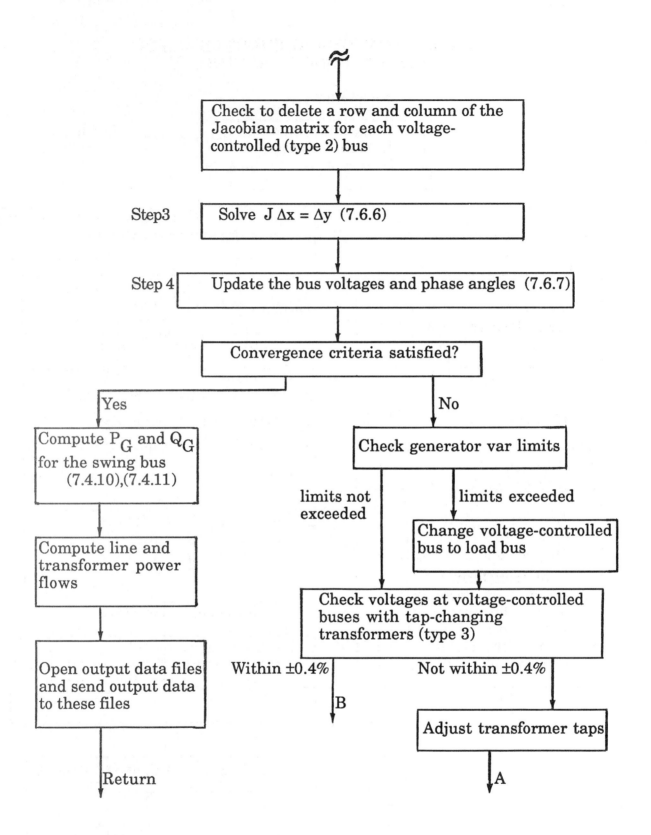

Check to delete a row and column of the Jacobian matrix for each voltage-controlled (type 2) bus

Step3   Solve J Δx = Δy (7.6.6)

Step 4   Update the bus voltages and phase angles (7.6.7)

Convergence criteria satisfied?

Yes

Compute P_G and Q_G for the swing bus (7.4.10),(7.4.11)

Compute line and transformer power flows

Open output data files and send output data to these files

Return

No

Check generator var limits

limits not exceeded

limits exceeded

Change voltage-controlled bus to load bus

Check voltages at voltage-controlled buses with tap-changing transformers (type 3)

Within ±0.4%

Not within ±0.4%

B

Adjust transformer taps

A

CHAP8   SYMMETRICAL SHORT CIRCUITS
(CHAP9   SHORT CIRCUITS)

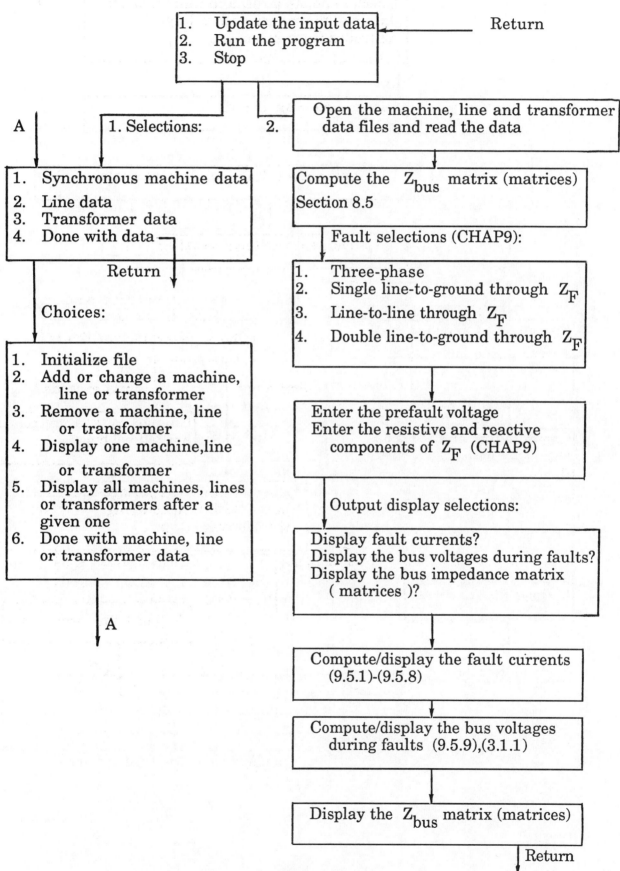

Selections:

1. Update the input data
2. Run the program
3. Stop

Return

A   1. Selections:

2. Open the machine, line and transformer data files and read the data

1. Synchronous machine data
2. Line data
3. Transformer data
4. Done with data

Return

Compute the Z_{bus} matrix (matrices) Section 8.5

Fault selections (CHAP9):

1. Three-phase
2. Single line-to-ground through Z_F
3. Line-to-line through Z_F
4. Double line-to-ground through Z_F

Choices:

1. Initialize file
2. Add or change a machine, line or transformer
3. Remove a machine, line or transformer
4. Display one machine,line or transformer
5. Display all machines, lines or transformers after a given one
6. Done with machine, line or transformer data

A

Enter the prefault voltage
Enter the resistive and reactive components of Z_F (CHAP9)

Output display selections:

Display fault currents?
Display the bus voltages during faults?
Display the bus impedance matrix (matrices)?

Compute/display the fault currents (9.5.1)-(9.5.8)

Compute/display the bus voltages during faults (9.5.9),(3.1.1)

Display the Z_{bus} matrix (matrices)

Return

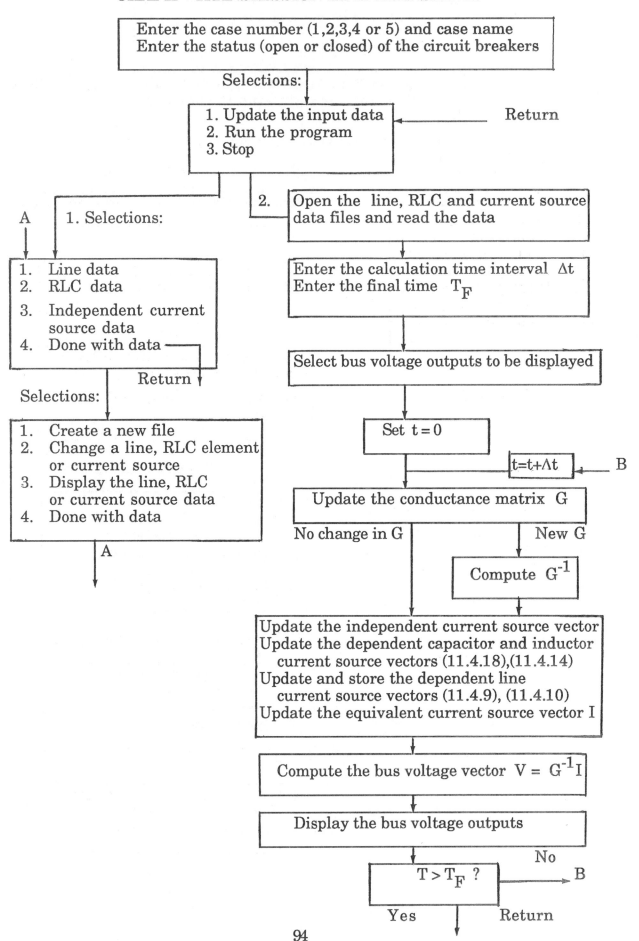

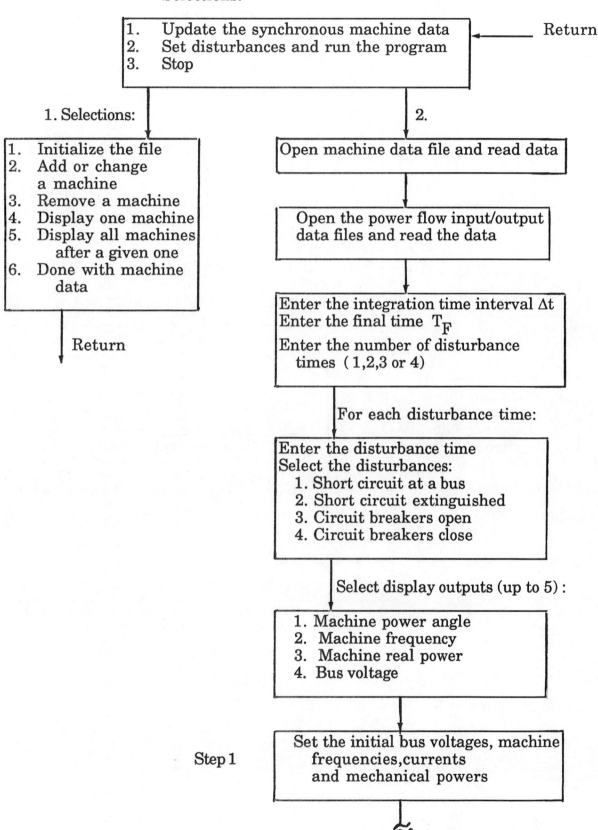

Selections:

1.   Update the synchronous machine data
2.   Set disturbances and run the program
3.   Stop

Return

1. Selections:

1.   Initialize the file
2.   Add or change   a machine
3.   Remove a machine
4.   Display one machine
5.   Display all machines   after a given one
6.   Done with machine   data

Return

2.

Open machine data file and read data

Open the power flow input/output
data files and read the data

Enter the integration time interval Δt
Enter the final time T$_F$
Enter the number of disturbance
   times (1,2,3 or 4)

For each disturbance time:

Enter the disturbance time
Select the disturbances:
   1. Short circuit at a bus
   2. Short circuit extinguished
   3. Circuit breakers open
   4. Circuit breakers close

Select display outputs (up to 5) :

1. Machine power angle
2. Machine frequency
3. Machine real power
4. Bus voltage

Step 1

Set the initial bus voltages, machine
   frequencies,currents
   and mechanical powers

≈

95

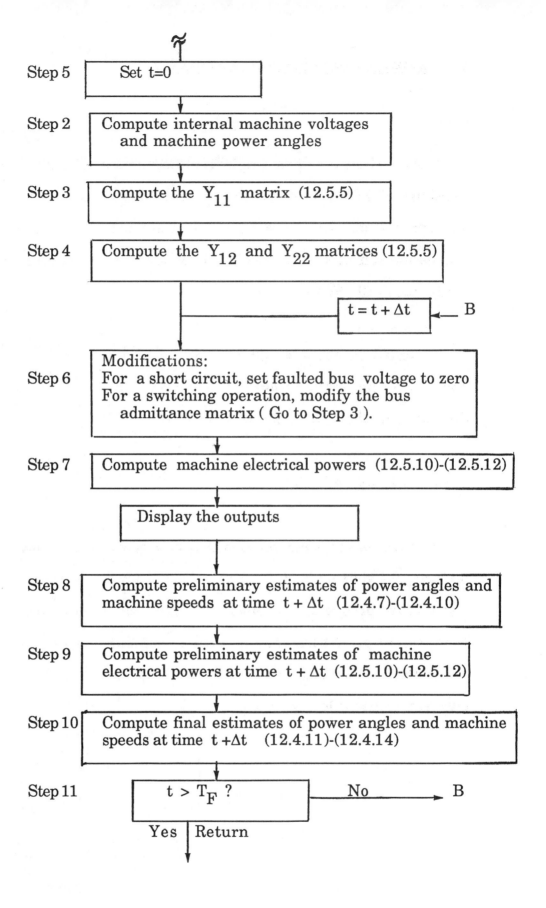

Step 5   Set t=0

Step 2   Compute internal machine voltages and machine power angles

Step 3   Compute the Y_{11} matrix (12.5.5)

Step 4   Compute the Y_{12} and Y_{22} matrices (12.5.5)

$t = t + \Delta t$ B

Step 6   Modifications:
For a short circuit, set faulted bus voltage to zero
For a switching operation, modify the bus admittance matrix (Go to Step 3).

Step 7   Compute machine electrical powers (12.5.10)-(12.5.12)

Display the outputs

Step 8   Compute preliminary estimates of power angles and machine speeds at time $t + \Delta t$ (12.4.7)-(12.4.10)

Step 9   Compute preliminary estimates of machine electrical powers at time $t + \Delta t$ (12.5.10)-(12.5.12)

Step 10   Compute final estimates of power angles and machine speeds at time $t + \Delta t$ (12.4.11)-(12.4.14)

Step 11 $t > T_F$?   No   B

Yes | Return

96

12. SAMPLE SINGLE-LINE DIAGRAMS

This section contains single-line diagrams and input data for original versions of IEEE 14-bus, 30-bus and 57-bus test systems, which are excerpted from the following publication: L.L. Freris and A.M. Sasson, "Investigation of the Load-Flow Problem", *IEEE Proceedings* , Vol 115, No.10, pp. 1459-1470, October, 1968.

Included in these single-line diagrams are synchronous condensers, three-winding transformers and static capacitors, which are briefly discussed in the following paragraphs.

Synchronous Condensers

A synchronous condenser, which is a synchronous motor operating at no-load, delivers or absorbs a variable amount of reactive power. In the power-flow program, the bus to which the condenser is connected is represented as a voltage-controlled (type 2) bus with zero real power generation. Synchronous condensers are identified by the letter C in the enclosed single-line diagrams.

Three-Winding Transformers

A three-winding transformer can be represented by three two-winding transformers connected in a T (or Y) circuit, as described in Section 4.6 of the text <u>Power System Analysis and Design with Personal Computer Applications</u> (Glover/Sarma). The equivalent impedances of the T circuit, computed from the transformer leakage reactances, are listed in the transmission line and transformer tables here.

Static Capacitors

A static capacitor, which is a capacitor bank connected from bus to ground, delivers reactive power. In the power-flow program, it can be represented by a load consisting of zero real power and negative reactive power $Q_L = - (V^2)(B_C)$ per unit, where V is the per-unit bus voltage and B_C is the per-unit susceptance of the static capacitor. At bus voltages near 1.0 per unit, this can be approximated by a constant negative reactive load $Q_L \approx -B_C$.

IEEE 14-BUS TEST SYSTEM

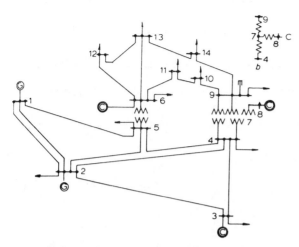

Figure 12a   Single-line diagram

Line designation	Resistance p.u.*	Reactance p.u.*	Line charging p.u.*
1–2	0·01938	0·05917	0·0264
1–5	0·05403	0·22304	0·0246
2–3	0·04699	0·19797	0·0219
2–4	0·05811	0·17632	0·0187
2·5	0·05695	0·17388	0·0170
3–4	0·06701	0·17103	0·0173
4–5	0·01335	0·04211	0·0064
4–7	0	0·20912	0
4–9	0	0·55618	0
5–6	0	0·25202	0
6–11	0·09498	0·19890	0
6–12	0·12291	0·25581	0
6–13	0·06615	0·13027	0
7–8	0	0·17615	0
7–9	0	0·11001	0
9–10	0·03181	0·08450	0
9–14	0·12711	0·27038	0
10–11	0·08205	0·19207	0
12–13	0·22092	0·19988	0
13–14	0·17093	0·34802	0

* Impedance and line-charging susceptance in p.u. on a 100000 kVA base
Line charging one-half of total charging of line

Table 12A   Transmission line & transformer data

Bus number	Starting bus voltage		Generation		Load	
	Magnitude p.u.	Phase angle deg	MW	MVAr	MW	MVAr
1*	1·06	0	0	0	0	0
2	1·0	0	40	0	21·7	12·7
3	1·0	0	0	0	94·2	19·0
4	1·0	0	0	0	47·8	−3·9
5	1·0	0	0	0	7·6	1·6
6	1·0	0	0	0	11·2	7·5
7	1·0	0	0	0	0	0
8	1·0	0	0	0	0	0
9	1·0	0	0	0	29·5	16·6
10	1·0	0	0	0	9·0	5·8
11	1·0	0	0	0	3·5	1·8
12	1·0	0	0	0	6·1	1·6
13	1·0	0	0	0	13·5	5·8
14	1·0	0	0	0	14·9	5·0

* Swing machine

Table 12B Bus data

Bus number	Voltage magnitude, p.u.	Minimum MVAr capability	Maximum MVAr capability
2	1·045	−40	50
3	1·010	0	40
6	1·070	−6	24
8	1·090	−6	24

Table 12C   Voltage-controlled (type 2) bus data

Transformer designation	Tap setting*
4–7	0·978
4–9	0·969
5–6	0·932

Table 12D   Tap settings of fixed-tap transformers

(The tap is on the 2nd bus number.)

Bus number	Susceptance** p.u.
9	0·19

** Susceptance in p.u. on a 100000 kVA base

Table 12E   Static capacitor data

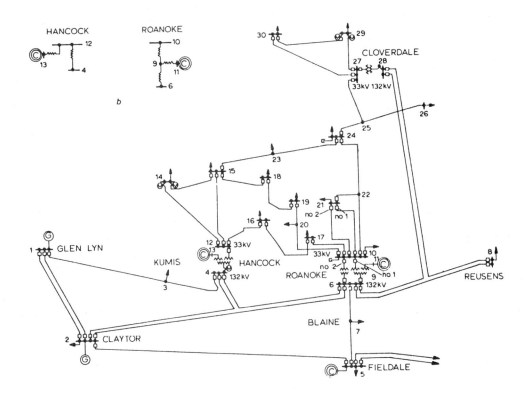

Figure 12b Single-line diagram

Bus number	Voltage magnitude p.u.	Minimum MVAr capability	Maximum MVAr capability
2	1·045	−40	50
5	1·01	−40	40
8	1·01	−10	40
11	1·082	−6	24
13	1·071	−6	24

Transformer desgination	Tap setting*
12–4	0·932
9–6	0·978
10–6	0·969
27–28	0·968

Table 12F   Voltage-controlled (type 2) bus data

Table 12G   Tap settings of fixed-tap transformers (The tap is on the $\underline{2}^{nd}$ bus number.

IEEE 30-BUS TEST SYSTEM

Bus number	Starting bus voltage		Generation		Load	
	Magnitude p.u.	Phase angle degrees	MW	MVAr	MW	MVAr
1*	1·06	0	0	0	0	0
2	1·0	0	40	0	21·7	12·7
3	1·0	0	0	0	2·4	1·2
4	1·0	0	0	0	7·6	1·6
5	1·0	0	0	0	94·2	19·0
6	1·0	0	0	0	0	0
7	1·0	0	0	0	22·8	10·9
8	1·0	0	0	0	30·0	30·0
9	1·0	0	0	0	0	0
10	1·0	0	0	0	5·8	2·0
11	1·0	0	0	0	0	0
12	1·0	0	0	0	11·2	7·5
13	1·0	0	0	0	0	0
14	1·0	0	0	0	6·2	1·6
15	1·0	0	0	0	8·2	2·5
16	1·0	0	0	0	3·5	1·8
17	1·0	0	0	0	9·0	5·8
18	1·0	0	0	0	3·2	0·9
19	1·0	0	0	0	9·5	3·4
20	1·0	0	0	0	2·2	0·7
21	1·0	0	0	0	17·5	11·2
22	1·0	0	0	0	0	0
23	1·0	0	0	0	3·2	1·6
24	1·0	0	0	0	8·7	6·7
25	1·0	0	0	0	0	0
26	1·0	0	0	0	3·5	2·3
27	1·0	0	0	0	0	0
28	1·0	0	0	0	0	0
29	1·0	0	0	0	2·4	0·9
30	1·0	0	0	0	10·6	1·9

* Swing machine

Table 12H   Bus data

Bus number	Susceptance* p.u.
10	0·19
24	0·043

* Susceptance in p.u. on a 100 000 kVA base

Table 12J   Static capacitor data

Line designation	Resistance p.u.*	Reactance p.u.*	Line charging p.u.*
1–2	0·0192	0·0575	0·0264
1–3	0·0452	0·1852	0·0204
2–4	0·0570	0·1737	0·0184
3–4	0·0132	0·0379	0·0042
2–5	0·0472	0·1983	0·0209
2–6	0·0581	0·1763	0·0187
4–6	0·0119	0·0414	0·0045
5–7	0·0460	0·1160	0·0102
6–7	0·0267	0·0820	0·0085
6–8	0·0120	0·0420	0·0045
6–9	0	0·2080	0
6–10	0	0·5560	0
9–11	0	0·2080	0
9–10	0	0·1100	0
4–12	0	0·2560	0
12–13	0	0·1400	0
12–14	0·1231	0·2559	0
12–15	0·0662	0·1304	0
12–16	0·0945	0·1987	0
14–15	0·2210	0·1997	0
16–17	0·0824	0·1923	0
15–18	0·1070	0·2185	0
18–19	0·0639	0·1292	0
19–20	0·0340	0·0680	0
10–20	0·0936	0·2090	0
10–17	0·0324	0·0845	0
10–21	0·0348	0·0749	0
10–22	0·0727	0·1499	0
21–22	0·0116	0·0236	0
15–23	0·1000	0·2020	0
22–24	0·1150	0·1790	0
23–24	0·1320	0·2700	0
24–25	0·1885	0·3292	0
25–26	0·2544	0·3800	0
25–27	0·1093	0·2087	0
27–28	0	0·3960	0
27–29	0·2198	0·4153	0
27–30	0·3202	0·6027	0
29–30	0·2399	0·4533	0
8–28	0·0636	0·2000	0·0214
6–28	0·0169	0·0599	0·0065

* Impedance and line-charging susceptance in p.u. on a 100 000 kVA base
Line charging one-half of total charging of line

Table 12I   Transmission line and transformer data

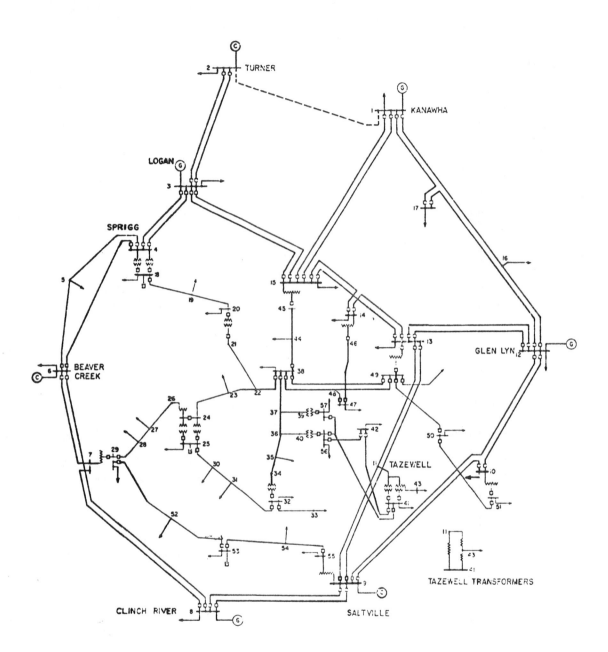

Figure 12c   Single-line diagram

Line designation	Resistance p.u.*	Reactance p.u.*	Line charging p.u.*
1–2	0·0083	0·0280	0·0645
2–3	0·0298	0·0850	0·0409
3–4	0·0112	0·0366	0·0190
4–5	0·0625	0·1320	0·0129
4–6	0·0430	0·1480	0·0174
6–7	0·0200	0·1020	0·0138
6–8	0·0339	0·1730	0·0235
8–9	0·0099	0·0505	0·0274
9–10	0·0369	0·1679	0·0220
9–11	0·0258	0·0848	0·0109
9–12	0·0648	0·2950	0·0386
9–13	0·0481	0·1580	0·0203
13–14	0·0132	0·0434	0·0055
13–15	0·0269	0·0869	0·0115
1–15	0·0178	0·0910	0·0494
1–16	0·0454	0·2060	0·0273
1–17	0·0238	0·1080	0·0143
3–15	0·0162	0·0530	0·0272
4–18	0	0·555	0
4–18	0	0·43	0
5–6	0·0302	0·0641	0·0062
7–8	0·0139	0·0712	0·0097
10–12	0·0277	0·1262	0·0164
11–13	0·0223	0·0732	0·0094
12–13	0·0178	0·0580	0·0302
12–16	0·0180	0·0813	0·0108
12–17	0·0397	0·1790	0·0238
14–15	0·0171	0·0547	0·0074
18–19	0·4610	0·6850	0
19–20	0·2830	0·4340	0
20–21	0	0·7767	0
21–22	0·0736	0·1170	0
22–23	0·0099	0·0152	0
23–24	0·1660	0·2560	0·0042
24–25	0	1·182	0
24–25	0	1·23	0
24–26	0	0·0473	0
26–27	0·1650	0·2540	0
27–28	0·0618	0·0954	0
28–29	0·0418	0·0587	0
7–29	0	0·0648	0
25–30	0·1350	0·2020	0
30–31	0·3260	0·4970	0
31–32	0·5070	0·7550	0
32–33	0·0392	0·0360	0
32–34	0	0·9530	0
34–35	0·0520	0·0780	0·0016
35–36	0·0430	0·0537	0·0008
36–37	0·0290	0·0366	0
37–38	0·0651	0·1009	0·0010
37–39	0·0239	0·0379	0
36–40	0·0300	0·0466	0
22–38	0·0192	0·0295	0
11–41	0	0·7490	0
41–42	0·2070	0·3520	0
41–43	0	0·4120	0
38–44	0·0289	0·0585	0·0010
15–45	0	0·1042	0
14–46	0	0·0735	0
46–47	0·0230	0·0680	0·0016
47–48	0·0182	0·0233	0
48–49	0·0834	0·1290	0·0024
49–50	0·0801	0·1280	0
50–51	0·1386	0·2200	0
10–51	0	0·0712	0
13–49	0	0·1910	0
29–52	0·1442	0·1870	0
52–53	0·0762	0·0984	0
53–54	0·1878	0·2320	0
54–55	0·1732	0·2265	0
11–43	0	0·1530	0
44–45	0·0624	0·1242	0·0020
40–56	0	1·1950	0
56–41	0·5530	0·5490	0
56–42	0·2125	0·3540	0
39–57	0	1·3550	0
57–56	0·1740	0·2600	0
38–49	0·1150	0·1770	0·0030
38–48	0·0312	0·0482	0
9–55	0	0·1205	0

* Impedance and line charging susceptance in p.u. on a 100000kVA base
Line charging: one-half of total charging of line

Bus number	Starting bus voltage		Generation		Load	
	Magnitude p.u.	Phase angle, deg	MW	MVAr	MW	MVAr
1 *	1·04	0	0	0	55·0	17·0
2	1·0	0	0	0	3·0	88·0
3	1·0	0	40	0	41·0	21·0
4	1·0	0	0	0	0	0
5	1·0	0	0	0	13·0	4·0
6	1·0	0	0	0	75·0	2·0
7	1·0	0	0	0	0	0
8	1·0	0	450	0	150·0	22·0
9	1·0	0	0	0	121·0	26·0
10	1·0	0	0	0	5·0	2·0
11	1·0	0	0	0	0	0
12	1·0	0	310	0	377·0	24·0
13	1·0	0	0	0	18·0	2·3
14	1·0	0	0	0	10·5	5·3
15	1·0	0	0	0	22·0	5·0
16	1·0	0	0	0	43·0	3·0
17	1·0	0	0	0	42·0	8·0
18	1·0	0	0	0	27·2	9·8
19	1·0	0	0	0	3·3	0·6
20	1·0	0	0	0	2·3	1·0
21	1·0	0	0	0	0	0
22	1·0	0	0	0	0	0
23	1·0	0	0	0	6·3	2·1
24	1·0	0	0	0	0	0
25	1·0	0	0	0	6·3	3·2
26	1·0	0	0	0	0	0
27	1·0	0	0	0	9·3	0·5
28	1·0	0	0	0	4·6	2·3
29	1·0	0	0	0	17·0	2·6
30	1·0	0	0	0	3·6	1·8
31	1·0	0	0	0	5·8	2·9
32	1·0	0	0	0	1·6	0·8
33	1·0	0	0	0	3·8	1·9
34	1·0	0	0	0	0	0
35	1·0	0	0	0	6·0	3·0
36	1·0	0	0	0	0	0
37	1·0	0	0	0	0	0
38	1·0	0	0	0	14·0	7·0
39	1·0	0	0	0	0	0
40	1·0	0	0	0	0	0
41	1·0	0	0	0	6·3	3·0
42	1·0	0	0	0	7·1	4·4
43	1·0	0	0	0	2·0	1·0
44	1·0	0	0	0	12·0	1·8
45	1·0	0	0	0	0	0
46	1·0	0	0	0	0	0
47	1·0	0	0	0	29·7	11·6
48	1·0	0	0	0	0	0
49	1·0	0	0	0	18·0	8·5
50	1·0	0	0	0	21·0	10·5
51	1·0	0	0	0	18·0	5·3
52	1·0	0	0	0	4·9	2·2
53	1·0	0	0	0	20·0	10·0
54	1·0	0	0	0	4·1	1·4
55	1·0	0	0	0	6·8	3·4
56	1·0	0	0	0	7·6	2·2
57	1·0	0	0	0	6·7	2·0

* Swing machine

Table 12L   Bus data

Table 12K   Transmission line and transformer data

Bus number	Voltage magnitude p.u.	Minimum MVAr capability	Maximum MVAr capability
2	1·01	−17	50
3	0·985	−10	60
6	0·98	−8	25
8	1·005	−140	200
9	0·98	−3	9
12	1·015	−50	155

Table 12M   Voltage-controlled (type 2) bus data

Transformer designation	Tap setting*
18–4	0·97
18–4	0·978
29–7	0·967
55–9	0·94
51–10	0·93
41–11	0·955
43–11	0·958
49–13	0·895
46–14	0·9
45–15	0·955
20–21	1·043
25–24	1·000
25–24	1·000
26–24	1·043
32–34	0·975
57–39	0·98
56–40	0·958

Bus number	Susceptance* p.u.
18	0·1
25	0·059
53	0·063

* Susceptance in p.u. on a 100 000 kVA base

Table 12O   Static capacitor data

Table 12N   Tap settings of fixed-tap transformers

(The tap is on the 2nd bus number.)